中文版《烟草制品管制：实验室检测能力建设》
© 中国科技出版传媒股份有限公司（科学出版社）2019

This translation was not created by the World Health Organization (WHO). WHO is not responsible for the content or accuracy of this translation. The original English edition *Tobacco Product Regulation: Building Laboratory Testing Capacity*. Geneva: World Health Organization; 2018. Licence: CC BY-NC-SA 3.0 IGO shall be the binding and authentic edition.

This translated work is available under https://creativecommons.org/licenses/by-nc-sa/3.0/igo.

世界卫生组织

烟草制品管制

实验室检测能力建设

主　译　胡清源
副主译　侯宏卫　陈　欢　刘　彤

科学出版社

北　京

内 容 简 介

本书提供了建设烟草制品实验室检测能力的备选方案，包括建立检测实验室的三种可能路径，即承包外部检测实验室，利用现有的内部检测实验室，以及开发烟草专用检测实验室。还为资源不足的国家开展烟草测试提供了切实可行的方法。

本书会引起吸烟与健康、烟草化学和公共卫生学等诸多领域研究人员的兴趣，可以为涉足烟草科学研究的科技工作者和烟草管制研究的决策者提供权威性参考，还对烟草企业的生产实践有重要的指导作用。

图书在版编目(CIP)数据

烟草制品管制：实验室检测能力建设/世界卫生组织编；胡清源主译. —北京：科学出版社，2019.3
书名原文：Tobacco Product Regulation: Building Laboratory Testing Capacity
ISBN 978-7-03-060804-8

I. ①烟⋯ II. ①世⋯ ②胡⋯ III. ①烟草制品 – 管制 – 实验室 – 检测 – 研究 IV. ①TS45

中国版本图书馆CIP数据核字(2019)第044289号

责任编辑：刘 冉 / 责任校对：严 娜
责任印制：吴兆东 / 封面设计：时代世启

科学出版社 出版
北京东黄城根北街16号
邮政编码：100717
http://www.sciencep.com

北京厚诚则铭印刷科技有限公司 印刷
科学出版社发行 各地新华书店经销
*
2019年3月第 一 版 开本：890×1240 A5
2019年10月第二次印刷 印张：5 7/8
字数：170 000
定价：98.00元
（如有印装质量问题，我社负责调换）

TOBACCO PRODUCT REGULATION
Building laboratory testing capacity

翻译委员会

主　译：胡清源

副主译：侯宏卫　陈　欢　刘　彤

译　者：胡清源　侯宏卫　陈　欢
　　　　刘　彤　韩书磊　付亚宁
　　　　王红娟

目 录

- 前言 1
- 致谢 4
- 术语表 5
- **第 1 章 国家监管机构背景下的测试** 7
  - 1.1 初步考虑 8
  - 1.2 如何使用数据 12
  - 1.3 确认烟草制品测试范围 17
  - 1.4 待分析物 19
  - 1.5 如何进行检测 21
  - 1.6 向监管机构提交数据 22
  - 1.7 涉及的成本 24
  - 1.8 实施 26
- **第 2 章 建立检测实验室的三种可能路径** 27
  - 2.1 承包外部检测实验室 28
  - 2.2 利用现有的内部检测实验室 30
  - 2.3 开发烟草专用检测实验室 31
- **第 3 章 承包外部检测实验室** 34
  - 3.1 实验室选择标准 34
  - 3.2 WHO 烟草实验室网络 41
  - 3.3 协议和法律与伦理问题 44
  - 3.4 样本量估计 46
  - 3.5 成本 47
  - 3.6 案例研究：加拿大 48

· i ·

3.7	循序渐进的过程	50
第 4 章	**利用现有的内部检测实验室**	**51**
4.1	要求（实验室设备、人员、总成本）——如何确定合适的实验室	53
4.2	认证	55
4.3	案例研究：新加坡	56
4.4	循序渐进的过程	57
第 5 章	**开发烟草专用检测实验室**	**59**
5.1	要求（基础设施、实验室设备、人员、总成本）	59
5.2	信息技术系统	61
5.3	数据验证	62
5.4	案例分析：CDC	63
5.5	循序渐进的过程	64
第 6 章	**资源：WHO TobLabNet 成员（标准、优势和程序）**	**66**
总结		**69**
参考文献		**70**
附录	**实验室内和实验室间验证**	**75**
A1.1	实验室内方法验证	75
A1.2	实验室间方法验证	77

Contents

Preface	79
Acknowledgements	82
Glossary	83

Chapter 1 Testing in the Context of a Country's Regulatory Authority 85

    1.1 Initial considerations 86
    1.2 How data can be used 91
    1.3 Identifying tobacco products to test 97
    1.4 Analytes to test 100
    1.5 How the tests are to be conducted 102
    1.6 Communicating data to regulators 104
    1.7 Covering costs 107
    1.8 Implementation 109

Chapter 2 Introduction to Three Possible Routes to a Testing Laboratory 110

    2.1 Contracting with an external laboratory 111
    2.2 Using an existing internal laboratory 113
    2.3 Developing a dedicated laboratory 115

Chapter 3 Contracting with an External Testing Laboratory 119

    3.1 Laboratory selection criteria 119
    3.2 WHO TobLabNet 127
    3.3 Agreements and legal/ethical issues 131
    3.4 Sample load estimates 135

3.5	Costs	136
3.6	Case study – Canada	137
3.7	Step-by-step process	139

Chapter 4 Using an Existing Internal Testing Laboratory —— 140

4.1	Requirements (laboratory equipment, staff, overall cost) – how to identify the right laboratory	143
4.2	Accreditation	146
4.3	Case study – Singapore	147
4.4	Step-by-step process	147

Chapter 5 Developing a Tobacco-exclusive Testing Laboratory —— 149

5.1	Requirements (infrastructure, laboratory equipment, staff, overall cost)	149
5.2	Information technology (IT) systems	152
5.3	Data verification	153
5.4	Case study–CDC	154
5.5	Step-by-step process	156

Chapter 6 Resources: WHO TobLabNet Membership (criteria, advantages, and procedures) —— 158

Summary —— 163

References —— 165

Appendix 1. Intra- and Inter-laboratory Validation —— 169

A1.1	Intra-laboratory method validation	169
A1.2	Inter-laboratory method validation	171
	References	172

前 言

众所周知，烟草的使用是一个重要的公众健康问题。然而，烟草制品是为数不多的在成分、设计特征和释放物方面几乎不受管制的公开消费品之一。大多数国家不愿在这一领域执行规定，部分原因是与烟草制品管制有关的技术复杂性。世界卫生组织成员国对收集有关烟草检测的资料和各国实验室能力建设的资源有很大需求[①]，特别是为了促进实施世界卫生组织《烟草控制框架公约》第9条和第10条。这将为监管机构和政策制定者提供有用、易懂的指南，指导他们如何检测烟草测品，检测什么产品，以及如何以一种有意义的方式使用检测数据来支持监管。

世界卫生组织《烟草控制框架公约》（WHO FCTC）反映了实验室检测的重要性。WHO FCTC 第9条规定了缔约方在烟草制品检测方面的义务，而第10条则规定了关于烟草制品成分和释放物的信息披露。产品信息披露有两种形式：①制造商向监管机构披露信息；②监管机构向公众披露信息。烟草制品检测生成支持这两种披露形式所需的数据。

---

① 世界卫生组织 2016 年 4 月在印度新德里举办了一个关于如何建立检测实验室的研讨会，与会者要求世界卫生组织编写一本关于实验室能力建设的手册。此外，世界卫生组织烟草实验室网络第六次会议（荷兰马斯特里赫特，2016 年 5 月 9~11 日）建议制定一份向各国政府和公众通报 WHO TobLabNet 活动的入门手册，以扩大成员数量，并在全球进行检测能力建设。

烟草制品管制
实验室检测能力建设

2006 年，WHO FCTC 第一次缔约方会议（COP1）设立了一个工作组，为实施公约第 9 条和第 10 条制定指南和建议（FCTC/COP1(15) 号决定）。第二次缔约方会议（COP2）延长了工作组的任务期限，并鼓励世界卫生组织无烟草行动组 (WHO TFI) 继续开展烟草制品管制工作 (FCTC/COP2(14) 号决定）。2010 年，第四次缔约方会议（COP4）提交的部分指南被通过。部分指南目前包含了降低烟草制品吸引力的法规建议。今后将拟订关于降低烟草制品致瘾性和有害性的建议。缔约方会议要求工作组继续其工作，在循序渐进的过程中详细阐述指南，并将有关致瘾性和有害性的最新情况提交以后的缔约方会议审议。

值得注意的是，与烟草行业的主张相反，这些指南是不可更改和有效的。部分指南所提倡的管制措施将被视为最低限度的要求，并不妨碍缔约方采取更全面的措施。

世界卫生组织一直支持成员国发展实验室能力。2004 年，世界卫生组织无烟草行动组（WHO TFI）发表了 WHO 烟草制品管制研究小组 (TobReg) 关于"提高实验室能力以促进 WHO FCTC 第 9 条和第 10 条实施，并指导烟草制品检测开展的指导原则"的建议[1]。该指导原则为打算发展这种能力的国家提供了建议，并帮助其实现这一目标。在这几年中，发展了新的知识，取得了新的进展，其中包括 2005 年建立的世界卫生组织烟草实验室网络 (TobLabNet) 和 2016 年建立的全球烟草管制论坛 (GTRF)。因此，应更新以前的文件，并为有意发展或取得烟草制品检测能力的国家提供切实的指导，以支持其监管机构。

本书提供了实验室能力建设的几种路径，包括开发检测实验室，利用现有的内部实验室，承包外部实验室，以及利用现有的支持机制，

前　言

包括但不限于世界卫生组织烟草实验室网络。最后，本书为实施烟草检测提供了切合实际、循序渐进的方法，甚至适用于在资源不足的国家建立检测设施。

致　　谢

本书主要贡献者为 David L. Ashley 博士（美国公共卫生署，已退休，佐治亚州立大学公共卫生学院环境卫生系兼职教授），以及世界卫生组织预防非传染性疾病部门的工作人员。

感谢 Nuan Ping Cheah 博士（新加坡卫生科学局医药、化妆品和卷烟测试实验室主任，世界卫生组织烟草实验室网络主席）、Ghazi Zaatari 博士（黎巴嫩贝鲁特美国大学病理与实验医学系教授和主席，烟草制品管制研究小组主席）和世界卫生组织《烟草控制框架公约》秘书处，审阅本书并提出有益建议。

本书的出版得到美国食品和药品监督管理局的资助（项目编号 RFA-FD-13-032）。

术 语 表

Accreditation	认可	来自独立机构的文件,证明一个实验室具有能够生成可被追踪和验证的可靠结果的体系
Accuracy	准确度	测量值与真值的接近程度
CDC	美国疾病控制和预防中心	
DAD	二极管阵列检测器	
FID	火焰离子化检测	
Firewall	防火墙	确保数据和信息受到保护,以使公众健康和商业利益分开,互不干扰的系统
GC	气相色谱	
HPLC	高效液相色谱	
Labstat	Labstat 股份有限公司	一个私营的商业化烟草分析实验室,位于加拿大安大略省的基奇纳市
LC	液相色谱	
MS	质谱	
MS/MS	串联质谱	
NCEH	美国国家环境健康中心	隶属美国疾病控制和预防中心
OSH	吸烟与健康办公室	隶属美国疾病控制和预防中心
PAHs	多环芳烃	多环芳香化合物,从两环(萘)到多环结构(如茚 [1,2,3-c,d] 芘)
Precision	精密度	使用相同方法重复测量同一样品时各测量值彼此的接近程度
Quality control	质量控制	评估系统是否在标准参数下持续运行的过程
Ruggedness	稳健性	分析系统能够承受偏离所定义的分析方法的能力
Selectivity	选择性	当一种物质确实不存在时,正确识别它不存在的能力
Sensitivity	灵敏度	测试方法在低水平下保证准确性和精密性的能力
TCD	热导检测器	
TFI	WHO 无烟草行动组	
TobLabNet	WHO 烟草实验室网络	
TobReg	WHO 烟草制品管制研究小组	
TSNAs	烟草特有亚硝胺	N-亚硝基降烟碱 (NNN),4-(N-甲基亚硝胺基)-1-(3-吡啶基)-1-丁酮 (NNK),N-亚硝基新烟草碱 (NAT),N-亚硝基假木贼碱 (NAB)
UV	紫外	
WHO	世界卫生组织	
WHO FCTC	世界卫生组织《烟草控制框架公约》	

第1章 国家监管机构背景下的测试

烟草制品检测是支持烟草控制和管制工作的有用工具，可以对人群健康产生明显的影响。烟草制品检测不会降低烟草中有害和致癌成分的含量，也不会减少烟草制品的使用，在烟草制品使用中，烟草制品的释放物会使使用者和非使用者接触到烟草制品中的有害化学物质。然而，有关设计、成分或释放物的数据将如何用于监管目的，如果对这一问题预先阐明目标和理由，这将是一个有效的工具。如何使用该工具是决定烟草制品检测是否有效的关键因素。

应该从一开始就理解，仅仅获取有关设计、成分或释放物的数据并不是要求烟草制品进行检测的充分理由。关键是国家要对如何使用这些信息构建合理的理由，因为测试很昂贵，即便制造商被要求完全资助这项工作，为了防止法律攻击或解决政府官员提出的合法问题，可靠的证据很重要。就本书而言，"制造商"一词指烟草制品制造商、进口商或其他隶属国家烟草制品监管机构并负责烟草制品销售的公司。

制定烟草制品检测方案的第一步是确定检测和报告的基础和理由。除了提供重要的证据之外，还应重点考虑使方案朝着利益最大化的方向发展，并确保提供的数据能够有效地支持国家烟草控制和管制。虽然一些很好的例子可以说明其他国家是如何实现产品测试需要的，但每个国家的经验不同。确定如何在国家层面使用信息是建立所有进一步行动的基础，必须首先考虑各国适合本国国情的需要。

根据政府监管机构要求烟草制品检测的方式，在烟草监管计划中使用实验室有两种方式。一种方式是由政府机构监督或对所有产品进行常规分析。在这种情况下，需求可以落实到位，使这些分析的成本由制造商完全承担（见 1.7 节），但这一努力将需要大量的行政工作，许多国家可能不想承担。另一种方式是，使用随机方案选择一系列品牌或子品牌的待分析物，由制造商提交数据，政府自己进行再分析，或采用其他机制来确保准确性，各国可以选择从测试到评估制造商提交数据的准确性都不直接参与。如上所述，可以要求制造商为测试提供经费。本书阐述这两种方法。

1.1 初步考虑

确定需要进行哪些测试的第一步是仔细评估政府要求进行测试的权力。每个国家的授权法律都不相同，在决定如何最好地使用将要获得的测试数据时，必须考虑到这些差异。对许多国家来说，世界卫生组织《烟草控制框架公约》（WHO FCTC）的条约义务可以指导国家烟草管制立法。WHO FCTC 第 9 条和第 10 条及其部分指南在制定国家烟草制品管制立法框架的同时，也列出了缔约方应采纳哪些条约的建议。

1.1.1 监管机构的设立原因

第一个需要考虑的问题是建立烟草制品监管机构的主要目的。原因如下：

- 烟草制品的使用对儿童健康的危害是被普遍关注的问题。虽然烟草使用造成的许多健康后果需要很长时间才能显现出来，但烟草制品成分和释放物对儿童有特殊吸引力，这可能是对其进行检测的最重要的原因。
- 其他成分或设计特性也是进行分析的重要原因，并将对检测能力作出早期决策。
- 立法可能是基于非使用者不接触有害化学成分的权利。在这种情况下，二手烟中有害释放物的测量可能是烟草制品检测中最关键的数据。在二手烟气中测量化学物质比在主流烟气中测量更具挑战性，因此，如果这是一个关键问题，应该考虑实验室进行这些测量的能力。
- 另一个可能的问题是烟草行业的虚假广告和其他营销声明。在这种情况下，检测能力能够对此类声明进行验证就很关键。
- 如果有人声称一种产品的风险较低，那么产品测试解决这些问题的能力应该是一个主要考虑因素。

还有一些问题可能是立法的主要考虑因素，在做出检测决策时也应该考虑这些问题。

1.1.2 立法的科学基础

另一个需要考虑的同样重要的方面是立法的科学基础。在决定何种实验室能力更必要时，重要的是要考虑最初的决定及其理由是否得到科学数据的有力支持，是否受到公共卫生政策关注的推动，或是否基于条约义务。如果立法基于科学数据，那么支持该立法的实验室数据可能是监管机构做出决定的关键辩护之一。在这种情况下，来自实验室测试的数据必须是无懈可击的。烟草监管机构必须确保向公众公开数据会加强法律证据。在大多数情况下，有一个强有力的科学依据来规范烟草制品，在最终如何使用这些数据方面，选择一个成熟、有经验和经过认证的实验室是不可或缺的。

1.1.3 公共卫生问题

从公共卫生的观点来看，监管机构应该考虑确定所关注的主要问题。这将因国家而异，重要的是监管机构要确定哪些问题是最重要的，并可以利用实验室数据加以解决。例如，使用卷烟会导致与烟草有关的最大的公众卫生问题吗？可能在许多国家都是如此，但另一些国家可能对其他烟草制品有更大的公共卫生关注，例如各种形式的无烟烟草、水烟、bidis、kreteks 或调味产品。如果这些其他形式是最大的公众卫生问题，那么检测卷烟可能是最直接的，但不是烟草制品检测帮助有效监管行动的最有效的选择。另一个公共卫生问题可能是新产品的引入。虽然新产品的总体影响最初可能不清楚，但烟草管制机构可能会被问及这些问题。提供基于科学答案的能力将证明检测能力的有用性。将检测资源集中在公共卫生关注的真正源头上，会证明需要进行检测的更好的理由，并且更容易实现公共卫生方面的真正改进。

烟草监管机构还应查明管制方案的真正目标，以及烟草制品检测如何提供帮助。具体目标可能是禁止某些产品进入市场，减少使用，或减少烟草制品使用造成的疾病和死亡。通常情况下，决策往往基于什么看上去更容易实现，而非是否符合烟草管制的总体目标。当烟草制品检测旨在支持主要的战略目标时，它就显示了价值，并会带来更好的结果。

1.1.4 公众与决策者的共鸣

同样重要的是要考虑最能引起公众和监管机构共鸣的数据,以及如何使这些数据变得可理解。这很重要,因为如上所述,烟草制品检测只有在与烟草控制或监管机构协调并被其使用时才有效。例如,如果成瘾是一个重大的公共卫生问题,那么测量烟草制品的主要致瘾性成分烟碱可能就是实验室能力的主要目标。如果产品的有害性及其对引发疾病的影响是主要问题,则将指向对有害和致癌物质的测量,如烟草特有亚硝胺(TSNAs)和多环芳烃(PAHs)的释放。如果烟草控制机构需要数据表明非吸烟者正暴露在二手烟气中,这可能是实现公共卫生目标的最有效方式。如果这些数据解决了公众和决策者清楚的公共卫生问题,或者可以明确表示,则这些数据更有可能被用于大幅改善公共卫生。

1.1.5 法律要求

最后,必须考虑到国家的法律要求及其可能对执行的影响。缺乏认证或不遵守标准可能会妨碍证据能力,或降低在法庭诉讼中附加在实验室报告上的权重。在选择进行烟草制品检测的实验室时,诸如监管链或参与实验室内验证(见附录)等问题是重要的考虑因素。烟草制造商很可能会使用高质量的实验室,而针对制造商的任何合规或执法程序,都需要符合科学来源的资质,才能具有可信度。政府机构在选择实验室以协助执法和其他程序时,应考虑有关的证据规则。他们还可以与司法部或其他法律行政部门合作,同时制定测试要求,以确保实验室数据足以作为法律程序的证据。

1.2　如何使用数据

实验室数据通过提供具有科学确定基础的事实陈述来避免意见性或传闻性声明。如果没有强有力的科学数据，监管行动的理由将更容易受到质疑和反驳，且采取的任何行动都不太可能实现其目标。虽然实验室数据不能保证成功，但科学数据加强了理性依据，增加了实现目标的可能性。

可用于减少烟草制品使用带来的疾病和死亡的执法机构因国家而异。但各国政府授予的权力既有风险，也有机遇。不幸的是，大多数烟草监管机构没有足够的资源来利用所有可能获取的数据。把机构扩展得太薄弱会降低实现目标的可能性。因此，至关重要的是，监管机构应评估利用现有权力和资源可以做些什么，以及如何最有效地利用它们。确定如何使用这些数据之前应该确定收集了哪些数据，以及对这些数据有哪些需求。

各国使用检测数据的一些可能方式包括：
- 制定产品标准；
- 限制广告声明；
- 教育公众；
- 为未来立法提供信息；
- 市场营销授权；
- 开发科学信息以支持研究；
- 制定生产标准。

1.2.1 产品标准

当一个国家具有监管机构时，制定产品标准可以成为减少烟草制品使用导致疾病和死亡的一种宝贵的监管工具。但由于这些标准非常有效，因此很可能会在法庭上受到挑战。为应对这些挑战，必须从有良好记录和同行评议的科学证据中获得支持。国家监管机构使用的产品标准包括那些针对烟草制品致瘾性和吸引力的标准。

- 2012年，巴西颁布了一项产品标准，禁止在巴西销售的卷烟和其他烟草制品中使用添加剂[2]。这是基于添加剂有鼓励年轻人使用产品，促使他们开始使用的影响。
- 2009年，加拿大颁布了一项禁令，禁止卷烟和雪茄中加入除薄荷醇外的其他香料[3]。这项行动的目的是"保护加拿大人的健康"，"保护青年人和其他人"，以及"提高公众对烟草使用危害的认识"。这项禁令最近扩展到包括薄荷醇[4]。
- 2009年，法国也通过了一项法律，限制在卷烟中添加香料，以减少青少年吸烟[5]。

所有这些标准都建立在实验室数据基础上，这些数据表明烟草制品中存在令人担忧的成分。产品标准的选择应该由上面讨论的国家层面的特定关键问题驱动。

1.2.2 营销与广告限制

一些国家已经限制营销与广告以降低产品的吸引力，可以限制直接营销（例如某些产品的低风险声明）或间接营销（例如在包装中使用颜色和图像增加吸引力）。例如，2001年，巴西第一个禁止在烟草制品标识上使用"轻"和"低焦油"等误导性词语[5]。虽然实验室数据本身不太可能足以支持采取行动限制营销与广告，但它可以作为评估相对风险声明和支持禁止或限制这些声明的行动的一个因素。

1.2.3 公众教育

公众教育可能是刚刚开始建立烟草管制方案的国家使用检测数据的最有效应用之一。一般来说，公众并没有相关科学知识。使用者和非使用者都不理解烟草制品的设计、成分和释放物如何影响他们的健康。许多人不明白，接触有害化学物质更多是烟草制品制造和使用过程造成的[6,7]。加深对产品检测信息的理解是一种有价值的方式，可以告知使用者，阻止其吸烟。但重要的是向公众提供的信息在科学上是正确的。对政府机构提供准确信息的可靠性的信任对于打击烟草制品制造商提供虚假信息至关重要。

公众教育可以成为一种有价值的工具，帮助使用者做出明智的选择，并使非使用者认识到与接触烟草制品释放物有关的危险。这种教育可以有许多不同的形式。加拿大的经验表明，公众对数字计算的理解很弱，会被吸烟机产生的关于产品的数字误导，因此发布《管制影响分析报告》，将《烟草制品信息规定》（TPIR）修正如下：

> 研究表明，目前的"有害释放物声明"格式显示了六种有害物质的一系列数值，烟草使用者通常不会注意到这一格式，很多人会感到困惑。拟议的法规将用四种基于文本的语句取代目前显示的数值，这些语句提供关于烟草烟气中有害物质的清晰、简明和易于理解的信息[8]。

另一方面,公众通常希望避免接触"化学物质",尤其是当这些物质与不良的健康结果有关时[9]。因此,虽然必须注意如何提供烟草制品中有害和致瘾性物质的资料,但应以有意义的方式向公众提供这些资料。

1.2.4 为未来立法提供信息

大多数国家在 WHO FCTC 公约义务方面的立法很可能并不全面。来自产品检测的数据为未来的立法提供了信息。未来的立法可以采取许多不同的形式,并应根据上文所讨论的烟草管制方案的总体目的仔细考虑。如果产品标准授权没有列入最初的立法,则烟草制品的检测可能是证明其价值的一个宝贵的数据来源。另一个新立法的例子可能是完全禁止在室内公共场所、工作场所和公共交通场所吸烟,就像许多国家近年来颁布的那样。由于室内禁烟很大程度上是出于对非吸烟者接触二手烟的担忧,因此这项工作的重点将放在非吸烟者接触的化学物质上。产品检测可以识别和量化这些非吸烟者接触的烟草制品所释放的有害和致瘾性化学物质。

1.2.5 通告或营销授权

一些国家在新产品进入市场时收到通告,一些监管机构在其国家立法或烟草控制法律中规定了通告要求。作为这一过程的一部分,可能要求烟草制造商提供详细的产品信息,包括以燃烧和未燃烧形式销售的烟草制品的成分、数量、有害性以及可能的副作用。利用检测对这些产品及其成分进行评估,是 WHO FCTC 授权的检测和报告义务的有价值的利用方式。

大多数国家在营销前没有授权来确定产品是否可以被授权销售。对于那些有授权的国家来说,这可能是一个非常强大的监管机构,但它需要大量的内部资源。当被授予这种权力时,产品检测是评估市场授权的一个重要因素。因为公司想被允许销售他们的产品,他们愿意提供大量必要的检测数据。为了正确地使用这些数据,监管机构必须有专门的科学资源和专业知识来评估制造商提供的数据。

1.2.6 科学信息开发

从烟草制品检测中产生的数据可以用来开发科学信息,这些信息对其他人(如研究人员)可能有价值,以便更好地了解该国烟草制品对疾病和死亡的影响。由于资源有限,与商业秘密有关的问题以及常规设计改变对烟草制品释放物的影响,研究人员很少能够全面评估他们的检测对象,即正在使用的烟草制品。如果可以从制造商获得这些信息并将其提供给研究人员和其他有关方,这将是改进对某一国家烟草制品使用的人类研究结果解释的一个有价值的工具。然后可以使用这些数据来支持上面列出的许多目的,包括制定未来的法规。

1.2.7 生产标准编写

生产标准是解决烟草制品使用所造成危害的另一个监管工具。检测数据可以识别生产原料中化学成分水平的变异性。例如,烟草特有亚硝胺(TSNAs)是烟草制品中最强的致癌物质[10,11]。但是传递给使用者的亚硝胺的水平高度依赖于生产中使用的原料烟草中的致癌物质的水平,这些水平根据烟草的不同有很大的不同[12]。通过设定生产中使用的化学成分或污染物（如重金属）的限量[13]，可以减少产品对使用者的暴露。但这些限量水平必须首先通过对烟草制品的有效检测来确定。

1.3 确认烟草制品测试范围

每个国家都有最需要监管的产品。如果评估检测结果的资源是有限的（通常是这样的），那么最好确定非常重要的产品，并将其作为优先检测产品。下面的讨论并不是要从检测需求中删除任何特定的产品，而是要建议监管机构考虑将哪些产品作为优先检测产品。

在评估将哪种产品列为最高优先级时，需要考虑三个因素：

（1）市场上最流行的烟草制品是什么？
（2）哪种烟草制品对使用者的危害最大？
（3）对哪些烟草制品进行管制是最可行的？

1.3.1 市场上最流行的烟草制品类型

为了对减少烟草使用造成的重大疾病和死亡影响，重要的是要解决有或可能有很大市场份额的产品类型。如果生产出来的卷烟只被一小部分人使用，即使大量减少使用也只会对公众健康产生轻微的影响。有几个数据来源可以用来评估不同类型烟草制品在一个国家的流行程度[14]。全球调查和类似的努力已确定使用不同类型烟草制品的人数。这些调查中还有许多根据性别和年龄划分使用情况。国与国之间产品使用的流行性差别很大。在印度尼西亚，kreteks（丁香卷烟）很受欢迎，但是这些产品在大多数其他国家只有很小的市场份额。无烟烟草在印度非常普遍,且使用的无烟产品种类非常多样化。另一个例子是薄荷醇卷烟的使用,这种卷烟在菲律宾等一些国家很普遍。因此,对一些国家来说,将监管产品检测的重点放在卷烟以外的产品上,可能是对资源的最佳利用。

1.3.2 对使用者最有害的烟草制品类型

烟草制品的有害性在产品种类内和种类之间是不同的。人们普遍认为,由于高浓度的有害和致癌化学物质被输送到肺部,传统产品(卷烟、雪茄、bidis、kreteks、水烟等)对使用者的危害很大。无烟烟草制品的多样性对监管和检测提出了进一步的挑战。但不同产品的有害性可能因具体的生产实践和使用者行为而异。一种无毒但经常使用的产品可能比毒性更强但很少使用的产品更令人担忧。在选择应重点关注的烟草制品时,监管者应考虑其市场中哪些产品对健康构成最严重的威胁。

1.3.3 最具监管可行性的烟草制品

在确定烟草制品时,第三个要考虑的因素是监管条件,以便将最初的产品检测需求集中在这些产品上。根据立法和政治环境的属性所赋予的特定权力,一些行动可能比其他行动更容易完成。如上所述,新产品被引入市场时,没有很大的市场份额,与有着强大利益相关者支持的成熟产品相比,可能是更可行的初始目标。一般来说,在有限数量的设备中生产且使用者不能对其进行重大修改的产品,比家庭手工业生产的产品更容易受到监管。当产品由成千上万的小制造商生产时,实行所需的检测可能非常具有挑战性,因此在建立新的检测和报告需求时,最初不应该把这些作为优先产品。监管机构应该考虑成功监管一个产业的可行性,这是优先化进程的非常普遍的部分。这种担忧的一个例子是印度的 bidis 制造业。Bidis 是一种手工卷制烟草制品,被广泛使用,给使用者带来了严重的健康问题。这些产品的大量生产作坊在私人住宅或非常小的商店里。在印度对这些小作坊产品实施检测要求将非常具有挑战性,可能不是优先检测的产品。在成功解决其他产品的问题之后,bidis 可能会成为一个目标。此外,使用者对烟草制品制造商的数据有效性进行篡改(例如添加石灰以增加游离烟碱水平)应作为确定产品是否纳入最高优先级进行测试和数据报告的部分考虑因素。

1.4 待分析物

一些国家已经制定了待分析物(用化学分析测量的化学物质)清单。加拿大是最先确定测量主流烟气、侧流烟气和全烟草分析物清单的国家之一[15]。2007 年,巴西也制定了一份设计特性、成分和释放物的待分析物清单[16]。2012 年,美国食品和药品监督管理局(FDA)公布了一份含有 93 种有害/潜在有害成分的烟草制品和烟气的清单[17]。

这些清单可以作为打算要求对烟草制品的设计特性、成分和释放物进行检测和报告的国家的起点。但是,决定应该检测哪些待分析物涉及几个因素,应该仔细考虑。

正如第 1.2 节所描述的,第一个因素是哪些待分析物能最好地满足数据如何使用的目的。应该使用一种基本原理来选择待分析物,将检测数据的结果与其应用联系起来。例如，如果监管机构计划向公众或决策者传达对致癌化学物的关心，则应明确选择 TSNAs 和多环芳烃（或以苯并 [a] 芘作为多环芳烃的代表）进行检测，因为它们是已知的致癌物，与烟草制品使用者罹患癌症有关。另外，为了制定产品标准而对重金属进行检测可能不是一个好的选择，因为烟草中的重金属主要是由种植烟草的土壤水平促成的，而不是制造过程造成的。重要的是要知道这在销售中的产品中是否是一个问题，以及是否可以使用一个标准来降低这些水平。如何使用检测数据是决定要测量哪种待分析物的关键因素。

一个同样重要的因素是，在第 1.3 节确定的产品中，哪些待分析物是最受关注的。在确定要在烟草制品成分和释放物中测定的待分析物时，应考虑某些产品的使用及其对健康的影响。例如，若关注烟草制品燃烧释放物，则应将一氧化碳作为待分析物。但是，由于无烟烟草在使用时不燃烧，所以在传统的无烟烟草中要求对一氧化碳进行测量是不必要的，也是不恰当的。重要的是所选的待分析物与要检测的产品相关。对仅由烟草燃烧产生的成分进行检测不适用于传统无烟烟草制品。

第三，对要检测的待分析物应该有可靠的测定方法。本书中，可靠性被认为包括适当的灵敏度和选择性以及适当检测范围内产生精确的和可重复结果的能力。当首先确定要在烟草制品成分和释放物中测量的待分析物清单时，已经建立广泛接受和灵敏的分析方法的待分析物应是最优先的，以便尽快获得结果。当检测程序更加成熟，并且已经获得了检测值时，可以在稍后添加没有建立方法的待分析物。

1.5 如何进行检测

如何对待分析物进行检测是开发实验室检测需求时的另一个重要考虑因素。不同国家以不同的方式处理这一问题，而且各国法律和可接受的要求也各有限制。一个问题是烟草制品成分或释放物中的待分析物是否允许使用不同的分析方法来测量，还是要求使用特定的方法。如加拿大所做的那样，要求使用特定的方法，具有一些明显的优点。烟草制品检测中会出现的问题之一是数据的可比性。由于准确度、灵敏度和选择性的差异，使用不同方法生成的数据并不总是具有可比性，即使它们应该是可比的。当使用相同的方法时，这些问题很大程度上被克服了，因为这些差异很大程度上被消除了。但是，由于实验室间在进行这些测量时存在差异，可能无法充分解决这些问题（见第 4.3 节）。通过参与实验室间的比较研究，可以在很大程度上解决内容实验室的差异。如果可能的话，确保可比性的更好方法是要求使用相同的方法在同一实验室进行分析，使数据的可比性最大化。WHO TobLabNet 已经开发并在全球验证了一些烟草制品重点成分和释放物的检测方法。

这种方法的缺点在于它的严格与死板。如果法律或规定要求采用同样的方法，则可能很难随着科学的进步采用更有效的新方法。所指定的方法随着时间的推移将跟不上新的、更灵敏、更可重复的方法的发展，这些方法可能有利于为公共卫生目的解释分析数据，至少在法律或规定更新之前是如此。此外，如果允许，要求使用单个实验室将无法产生必要的竞争，这种竞争可以降低成本，并通过启动其他实验室的开发来鼓励开发其他的检测能力。如果允许，在接受数据之前，必须充分评估方法和/或实验室的选择、方法的可靠性及进行测试的实验室的可靠性（见第 4.2、4.3 和 5.3 节）。这需要大量的努力和专家的建议。要求测量已知的标准物质可以有所帮助。在决定采用哪种方法时，监管机构应该折中考虑这些问题。

1.6　向监管机构提交数据

有关提交给监管机构的数据的要求，与选择待分析物以及如何对其进行检测同样重要。监管机构需要了解如何进行测量，以便评估数据的质量和可比性并采取适当的行动。仅仅分析结果，而不考虑它们的意义，是有限的使用。

监管机构必须决定对每个品牌/子品牌报告数据的频率。可接受的频率可能是每年两次到每两年一次。更频繁的测量有助于评估生产运行之间的差异，但增加了成本和收集、编译整理和评估所提供数据所需的资源。在最终确定需求之前，监管机构应该权衡这些因素。

有一些报告需求是显而易见的。必须提供测试产品的无可争议的标识，这包括可以识别出特定产品的品牌/子品牌的信息。至少，子品牌信息应包括：

- 物品的尺寸[②]（卷烟的长度和直径）；
- 物品的数量或包装的大小（例如，20 支卷烟，3 盎司[*]无烟

---

② 物品是指消费者使用的特定产品。例如，对于卷烟来说，物品就是真正燃烧的烟棒。

* 1 盎司 =29.57 cm^3

烟草）；
- 添加的配料，包括香料（如薄荷醇、草莓、薄荷）；
- 无烟烟草截断尺寸（如切断的长度）；
- 卷烟的通风水平；
- 制造商或消费者用以区分同一品牌产品的任何其他指示物。

同样重要的是，测定的水平与测量单位（例如 mg/支卷烟、mg/g 烟草）一起报告。最后，应报告所作的所有分析决定、重复次数和统计上接受的数据之间的总平均数。重要的是，所有的结果，甚至那些被剔除的结果，都要连同被剔除的原因一起包括在内，以便能够正确地评估所报告的数据。监管机构还应该指定应该报告的有效数字位数（通常是 3）。当样品之间可能存在显著差异时（例如 3.12 mg/g 和 3.45 mg/g），如果要求的有效数字太少（例如 3 mg/g 和 3 mg/g），则导致得不到有效结果。

高度推荐建立分析测量质量的其他支持信息。向监管机构提交的报告应包括用于进行分析测量的方法和方法验证参数（见附录）。为了正确评估数据的可靠性，并了解是否可以将其与报告的其他数据和提交给监管机构的先前数据进行比较，或在同行评审的文献中进行比较，有必要了解所使用的方法及其准确度、可重复性、灵敏度和选择性，只有这样才能做出适当的比较。向监管机构报告的有助于证明报告结果质量的其他数据应包括质量控制结果，表明在进行测量时分析系统运行正常，以及已知标准材料的水平。已单独确定其测量水平的样品可以用来评价所报告的水平是否与科学上公认的结果一致，以及对未知样品所报告的结果是否有效。

包含关于如何选择检测样品的信息也很重要。此外，必须提供样品所在的位置（例如生产线、仓库、零售地点）以及样品所经历的运输和存储条件,因为某些待分析物在一定的存储条件下会发生变化。例如,在某些存储条件下,某些烟草制品中的 TSNAs 含量会上升[18]。此外,在存储过程中 pH 值的改变导致无烟烟草游离烟碱水平的变化[19]。在这种情况下,不适合对不同条件下不同储存时间的样品进行分析结果的比较。为了减少偏差,应该指定样本的选择方法。这可能包括对放置在同一仓库的样本进行随机取样、对来自多个生产线的取样、在零售场所盲选样本或其他方法。设计随机化方案,应使所选择的样本能够代表市场上销售的产品。这必须确保样品不是专门为分析测试而制造和取样的,而是代表了销售给消费者的产品。

1.7 涉及的成本

监管机构和政府不应负担检测和报告的费用。制造商应该承担所有成本作为经营和进入市场的条件。这是大多数行业(食品、化妆品、药品)的标准做法,也应该适用于烟草制品制造商。可以通过制造商和检测实验室之间的直接交易来解决检测成本的问题。监管机构没有必要充当中介,尽管其可能会这样做,从制造商那里接收样品,然后将它们发放出去进行检测。这给监管机构带来了沉重的后勤负担,由制造商承担最好。但为了确保数据的准确性和完整性,应采取本书所述的其他保障措施。

监管机构在参与检测项目时可能会产生一些费用,包括评估报告的数据、监督检测和报告系统、执行和分析以检查报告结果的真实性。各国可以使用一些机制来确保这些成本也由制造商承担。

第1章　国家监管机构背景下的测试

一些国家可能会选择向制造商征收用户费用，以支付政府的监管费用，并允许烟草制品销售。用户费用可以根据特定制造商在市场上的品牌/子品牌数量，或者特定产品的市场份额而定。应该为监管机构的运作设定一定的数额，可以对其进行细分，以便每个制造商支付适当的份额。不应该制定用户费用总额，导致在产品使用减少的情况下降低用户费用。相反，如果流行度降低，每一种产品的用户费用应该增加。这会起到其他作用，即如果流行度降低，每一种产品的价格将会上升。

非常重要的是，用户费用结构不要求监管机构允许营销或鼓励烟草制品销售的增加。例如，用户费用不应该基于销售的产品总数，随着流行度的增加，用户费用也会增加。这可能会引起监管机构的利益冲突。用户费用体系的设计应该鼓励减少烟草制品的销售，或者至少不会产生影响。它不能被设计成鼓励烟草制品整体销售增长的方式。当监管机构评估是否允许产品营销时，无论是否允许营销，基于此决定的用户费用的支付必须是相同的。用户费用不应基于允许产品营销的积极决策。

罚款可以为监管活动提供另一种收入来源，尽管它不应该是唯一的资金来源。例如，对非法或不合规行为（包括不报告）进行罚款。这种方法鼓励制造商遵守规定。与销售比例无关的商业活动的固定费用是另一个可能弥补成本的收入来源。

在使用这些融资机制时，应在制造商和执行监管活动的机构之间建立防火墙，以防止制造商有不正当的行为。这可以通过要求制造商向国家财政部支付适当的资金，并由监管机构获得适当的拨款来实现。但各国政府往往在寻找资金来源，以支持各种活动，因此，任何机制都必须由法律明确规定，并确保监管活动持续、确定和适当的资金来源。

1.8 实　　施

在考虑如何制定烟草制品检测方案时,一个重要的问题,是立即实施所有措施,还是循序渐进。在可能的情况下,一般建议采用循序渐进的方法。这使方案能够更快地启动,因为可以采取渐进步骤来处理明显的需求,而不是预测未来的每一种可能性。此外,它还允许监管机构从最初的错误中吸取教训并进行调整。如果采取一种"一劳永逸"的方式,那么改变方向可能会变得相当困难,以至于最初的决定可能会阻碍该计划走向未来。

在某些情况下,可能需要采取"一劳永逸"的办法。如果政治条件造成渐进的方法是不可能的,那么监管机构就可能被要求立即开始行动。虽然这种情况是可能的,但是这种方法有很大的风险。由于机构可能遇到困难,在制定测试方案之前和制定测试方案过程中更需要与专家进行仔细协商。关于建立实验室能力的各种方法的详细信息在接下来的四章中给出。有关 TobLabNet 提供的专业知识的讨论,请参见第 5.2 节。

第2章 建立检测实验室的三种可能路径

运行一个实验室并保持必要的实验室测量质量是昂贵的和资源密集的。烟草制品检测实验室需要有经验的工作人员，他们在大量严格审查且具挑战性的分析测量方面有成功经验。检测实验室需要有效的实验室信息管理系统，能够有效地处理、评估和存储大量数据。这些要求与典型的研究型实验室的要求不同，因为工作性质和强度不同。实验室需要昂贵的分析设备，必须定期维护、维修和更换。现代实验室仪器是非常复杂的设备，需要特别的专业知识来确保它们能够正常工作并符合规格。

实验室必须保持每天的分析可靠性，以便所有的结果都是一致和准确的。实验室需要外部认证和质量监控，以证明结果的质量和可靠性，并在受到严格审查时证明其精确性。要做到这一切，就必须保证对任何实验室的资金和人员资源都有定期和充分的支持，以维持其检测能力。能力是随着时间而发展的，必须加以维护，以便在需要时可以依赖。我们提出了烟草制品设计特性、成分和释放物检测实验室能力建设的三种方法。

本章总结了这些方法，后面的章节提供了关于这种能力的详细信息。

2.1　承包外部检测实验室

世界上有几个与烟草行业无关的经验丰富、独立的烟草检测实验室。为了鼓励这类实验室的发展，并更好地确保测量的质量和一致性，建立了烟草实验室网络（TobLabNet）。诸如此类的实验室一般有两种形式：独立的商业烟草制品检测实验室以及政府拥有和经营的烟草制品检测实验室。

如果独立的商业烟草制品检测实验室已经运行了很长一段时间，那么它们就具有一定的优势。它们应该已经具有经过充分测试和验证的能力，并具有参与内部实验室比较的经验。它们已经有了经验丰富的科学家和技术人员以及进行一系列分析所需的设备。它们习惯于在合同的基础上进行测试和报告，并准备在这些条件下提供测试结果。它们已经开发了 IT 系统，应该已经参与了外部质量保证计划。对于准备进行检测的监管机构来说，这些实验室可以快速响应并及时提供结果。但一般来说，它们仅限于自己的测试能力范围。它们有可能开发新的测试能力，但需要时间，而且需要保证开发和验证一种新方法对其商业有利。因此，针对国家特定的检测可能不是其当务之急。

一些独立的商业烟草制品检测实验室也在合同基础上为烟草行业进行测试。这可能涉及一些国家是否遵守 WHO FCTC 第 5.3 条。在这些情况下，各国应要求建立防火墙[③]，以保护公众健康不受商业

---

③　一种确保数据和信息受到保护的系统，使公共卫生和商业利益相互独立，互不干扰。

利益影响，并确保结果的机密性和独立性。为行业和监管机构进行产品检测的实验室不应被直接拒绝，而应进行评估，以确保其完整性、无偏见和保密性。

还有大量政府所属的烟草制品检测实验室。TobLabNet 继续成功地与世界各地的多个政府所有烟草检测实验室合作，鼓励其发展分析检测能力，并提供用于实验室内验证的机制（见附录）。因此，那些非常有效和可靠的政府实验室了解烟草制品管制检测的重要性和目标，并面临着与开始新方案时类似的挑战。这些实验室与独立的商业烟草制品检测实验室有许多相同的优势，包括经验、IT 系统、质量保证计划和已建立的能力。此外，对于一个刚刚启动项目的国家来说，与其他国家的监管机构合作可能具有很大帮助。

例如，这种相互作用可以在新的方案和成熟方案之间提供一种自然的协商关系。如果一个新的监管机构做出了不恰当的决定（例如，在释放物方面检测了不正确的分析物），政府机构更有可能提供建议，而不是进行不恰当的测试。这是新的烟草管制计划的一大优势。与政府现有实验室合作的最大缺点是，它们有自己的法定要求和优先事项，所以不太可能在预期的时间内进行测试。它们可能会因为其他的优先事项而被推迟，其管理层很可能会认为自己的优先事项应优先安排。

外部实验室的第三种选择是烟草制品制造商自己拥有和运行的实验室。在任何情况下都应避免这种情况，因为有内在的利益冲突。

2.2 利用现有的内部检测实验室

对于这项讨论，我们假设一个国家有一个经验丰富的实验室，该实验室已经在为其他目的进行测试。例如，该国可能进行环境或药物测试，并具有以前进行合规检测和报告的经验。这种方法有一些优点和缺点。

对于一个已经在检测其他消费品的实验室来说，有发展内部检测烟草制品能力的基础。在开发一个以前没有的有效实验室的过程中，最大的挑战之一是雇佣具有宝贵专业知识的员工，他们了解如何进行有效、合法的测量。此外，可用于烟草制品检测的实验室设备、IT系统和质量保证方案大部分已经到位。这种方法将比从零开始建立一个烟草检测实验室更经济、更快捷。使用这种方法可能还有其他优点。如果目前的实验室资金不足，无法达到预期的效果，烟草检测的额外资金可能会有所帮助。这对那些往往由有限的政府资金提供资助的实验室来说是一个主要优势。

不利的方面是，烟草检测可能需要新的设备和专业知识。例如，吸烟机及其使用方面的专门知识仅限于燃烧烟草制品的检测；环境或药物测试实验室不会有这种设备或经验。因此，获得这种能力仍然需要大量的启动时间和成本。但如果计划正确，有不需要这种能力的特定优先项目（例如卷烟烟草含量或无烟烟草检测），这可能是实验室开发的第二阶段。此外，如上所述，在现有的实验室中发展烟草制品检测能力可能会导致优先冲突。例如，一个药物检测实验室可能已经分配了充分的工作人员和设备。实验室很少有大量的过剩产能。因此，

第 2 章 建立检测实验室的三种可能路径

当两项计划都需要迅速获得结果时,自然就会发生冲突。明智的做法是在达成最后协议之前明确说明如何处理这些冲突。

如果有可以增加烟草制品检测能力的研究实验室,请注意工作的性质和科学方法是不同的。为符合规定或报告目的而测试产品的实验室的工作具有与研究实验室不同的要求。一般来说,研究实验室需要增加 IT 基础设施,建立更健全的质量保证系统,寻求认证,并准备提供法律证据,以便成功成为一个合格的实验室。相比之下,现有的内部符合规定的实验室应该已经习惯于生成可用于符合规定或法律目的的结果。第 4.3 节描述了烟草检测实验室使用其他设施来支持额外的烟草制品检测能力的案例研究。

2.3 开发烟草专用检测实验室

要考虑的最后一个选项是在不共享资源的情况下,建立一个专门的烟草制品检测政府实验室。这种方法有相当大的优点,拥有一个专门的检测实验室意味着检测能力和开发新的检测能力的重点将由烟草管制重点驱动。因此,现有能力的使用优先级可以由单个管理结构来确定。也可利用过剩产能作为增加额外收入的手段来支持实验室的运作。

另一方面,建立一个能够产生完全可靠结果的实验室需要投入大量时间、资金和人力资源。如果有现有的实验室设施,或设施和工作人员可重新分配到新的任务中,可能会有所缓解。如果没有,可能需要建造或改造物质结构。实验室需要特殊的空气处理、电力需求(如不间断电源)及其他通常不在办公室、零售店或商业建筑中使用的物理设施。这可能需要建造新的设施或改造现有设施。获得足够支持来发展一个需要多年建设和装备的实验室是有挑战性的，特别是在政府有其他优先事项的时候。若想在政府决策者中保持对测试能力的支持，可能需要提供数据来证明这些重大投资的价值，拖延可能导致失去支持。虽然这肯定是一个可行的选择，但有几个国家在尝试这种方法时失败了。他们的失败很大程度上是建设进度的拖延和政府管理的改变而造成的。

表1总结了上面描述的这些方法的优点和缺点，并在后面的章节中详细描述。

表1　开发实验室能力的方法的优缺点

实验室类型	优点	缺点
外部实验室——商业	• 不需要内部测量专业知识 • 启动成本较低 • 比开发新功能启动更快 • 广泛的功能 • 立即接触有经验的科学家 • 确认结果的有效性	• 可用性不能保证 • 必须定期评估可靠性 • 从长远来看，可能会更贵 • 可能仅限于实验室中可用的方法 • 开发新方法的灵活性较低
外部实验室——政府	• 不需要开发新的内部专业知识 • 启动成本较低 • 比开发新功能启动更快 • 立即接触有经验的科学家 • 确认结果的有效性 • 鼓励与其他国家的监管机构协商	• 产生结果可能会延迟 • 必须定期评估可靠性 • 从长远来看，可能会更贵 • 开发新方法的灵活性较低 • 可能对能力有限制
现有的内部实验室	• 需要一些专业知识 • 通过资源共享提高整体效率 • 启动成本较低 • 比创建新实验室启动更快 • 已有可靠的IT和质量体系 • 可能有助于稳定两种项目的资金	• 可用性很大程度上取决于其他优先事项 • 必须购买一些烟草专用设备（如吸烟机） • 必须发展一些专业技能（如吸烟机操作） • 需要一些启动时间和成本
专用的内部实验室	• 保证可用性 • 可以通过外部工作产生收入 • 可以根据需要开发能力 • 优先级的完全灵活性	• 拓展能力代价高 • 启动成本高 • 可能需要搭建设施 • 需要大量的启动时间 • 必须获得政府支持 • 政府支持和资金水平可能会波动

第2章 建立检测实验室的三种可能路径

对于开始检测计划的国家,建议从经验丰富的外部实验室开始(见图1)。有几个目的,其中最重要的三个目的如下:

(1)允许在有强有力的支持的情况下快速启动检测。

(2)快速提供有助于支持长期开展烟草制品检测基本理论的数据。

(3)减轻了监管机构对实验室开发和数据质量的负担,使其能够专注于其他关键问题。

如果有大量的样品和有力支持,监管机构可以考虑与国有经营的检测实验室达成协议进行烟草检测。如果维持外部合约,将允许不损失能力而是扩大容量的逐步过渡。最后,如果样品吞吐量支持,可以考虑建设和装备一个专门的国家烟草检测实验室。但是,只有在有足够数量的检测和行政支持持续强大的情况下,才应该这样做。如上所述,直接从不具备能力发展到专门的国家烟草制品检测实验室并没有被证明是可行的方法。

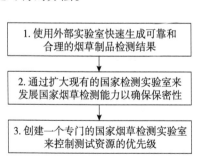

图1 建议构建测试能力的方法

第3章 承包外部检测实验室

3.1 实验室选择标准

在确定要与之签约的外部实验室时,有几个重要的考虑因素。在做出最后决定时,对每一项措施的重点会因国家而异,但所有这些都要加以考虑。在做最后决定之前,先调查一下这些考虑因素是值得的。

3.1.1 烟草分析经验

这是至关重要的,可能也是最重要的因素。如上所述,与外部实验室合作的主要原因是基于在进行烟草制品分析方面的特殊经验。选择经验有限的实验室与使用外部实验室的优势相矛盾。如前所述,烟草分析有它自己的要求、程序、设备和标准,一个在这一领域有多年经验的实验室和显示可追溯的记录将大大减少启动成本和时间。

3.1.2 设备性能

分析化学不断发展以开发准确度、可重复性、选择性和灵敏度更优的过程和设备。虽然当待分析物含量高时,提高灵敏度是不必要的,但烟草制品中的许多成分和释放量都处于挑战检测和精确量化能力的水平。现代分析设备正在稳步提高检测和量化限制,这些改进对于回答特定的烟草管制问题可能是至关重要的。对外部实验室的评估应包括对分析仪器的广泛性和复杂性的评估。为了满足当前的需要，实验室至少应包括后文表3所列的设备。各国还应考虑表4所列的设备以及可能用于特定国家所需的其他设备。任何实验室都需要考虑必须包括气相色谱（GC）-火焰离子化检测（FID）和液相色谱（LC）-二极管阵列检测器（DAD），如果同时还配备液相色谱串联质谱则更有优势。此外，自动化样品制备设备的使用程度将减少人为误差。预期烟草制造商能够配置现有的最先进的仪器设备，当在符合性、执行或法律规定中比较数据时，最好知道所使用的仪器对用于特定用途来说是最先进的仪器。所作决定应尽可能以最先进的分析为依据。

3.1.3 人员素质

有经验、有能力的员工是实验室最宝贵的财富。虽然可以购买先进的设备，但必须对工作人员进行培训，并需要时间积累经验。有经验的实验室分析人员和仪器操作人员对于任何商品进行可靠的分析测量都是必要的，特别是对烟草制品，因为有特殊的要求。在评价一个实验室时，应当对工作人员执行分析测量的次数和时间长短做出评价，特别是对烟草制品进行分析时。此外，定期培训新的和有经验的工作人员的制度应确保他们在开始独立分析之前能够胜任，并确保他们能跟上该领域的进展。

3.1.4 能力范围的广泛性

即使最初的分析范围是有限的，但可能在某个时候，情况会发生变化,将需要进行新的测试。一个只有有限能力的实验室可能不能在需要进行监管时进行新的分析。在这种情况下,它们需要大量的启动时间来开发能力并验证新的测试。如果这些能力已经存在,那么启动时间是最少的。此外,能力范围更广的实验室可能具有更高的质量。有能力的科学家总是在寻找新的挑战或改进方法。一个允许员工在工作中成长的实验室会吸引并留住更优秀的科学家。

3.1.5 足够的能力

为监管目的,测试结果必须具有准确度、可重复性、灵敏度和选择性(见第 4.3 节)。但是,它们也必须是及时的。如果实验室不能在需要时提供结果,关键时机可能过去,这些数据可能不再相关或有大影响。能力不仅是当所有系统都按预期运行时的样品通量,而且还应具有灵活性。例如,只有一台吸烟机的烟草烟气检测实验室,如果正在维修,就不能处理燃烧产品的释放物样品。当发生意外事件时,关键设备和工作人员备份设施可以继续发挥作用。能力应该包含在实验室的整体评估中。

3.1.6 已证明的可追溯记录

先进的设备、训练有素的工作人员和足够的能力不能确保实验室能够在最后期限内生成和报告可靠的结果。及时性也可以作为机构文化和管理有效性的指标。实验室应该能够提供参考或记录,表明可靠的结果可以定期生成和报告。如果可能的话,有必要获取之前的客户名单,并随机联系他们中的一些人,以评估实验室产生结果的能力是否正如所承诺的那样。

第 3 章 承包外部检测实验室

3.1.7 认证

认证是由一个独立机构提供的证书,该证书显示实验室有相应的体系,使其能够产生可充分跟踪和验证的可靠结果。有一些国际和国家实验室认可机构和认可标准,如 ISO 17025,有效地执行了这一功能(见第 4.2 节)。为合规和执行目的而产生结果的任何实验室都应获得认证。任何没有认证的实验室都不应该被考虑。即便如此,重要的是,仅凭认证不能保证有效的结果。实验室有可能获得认证,却不能产生适当的结果。

3.1.8 扩展新方法的能力

如上所述,有时需要进行以前没有预料到的专门分析,而且没有现有的方法。一个很好的例子是引入一种新的产品类型或可能产生新的释放物的改良产品。在这种情况下,实验室可能需要在相对短的时间内开发和验证一种新方法。实验室应该能够从以前的实例中给出实施这一过程的例子。由于这可能需要大量的开发工作,因此对新方法进行分析可能比常规检测成本要高。

3.1.9 信息技术系统

准确度是分析测试的基本要求。通过使用训练有素和经验丰富的工作人员,采用正确的分析方法和仪器,可以最大限度地提高准确度。但是,只要分析过程涉及数据的处理,就会出现错误。当数字被手动转录时就是一个特别的挑战。手工录入的数字越少,错误就越少。实验室信息管理系统(LIMS)在整个实验室检测中都很常见，被认为是检测和报告的必要条件。一个没有 LIMS 的实验室不应纳入进一步的考虑。自动化程度越高，发生人为错误的概率就越低，但是还需要一个检查 LIMS 是否正常工作的过程。从样品收据到报告准备都可跟踪样品和数据处理的 LIMS 是非常理想的。但是，检查也需要纳入质量保证方案，定期验证这些系统。

3.1.10 质量保证方案

质量保证将明确的质量控制计划与全面的质量意识相结合。质量控制评估体系是否在标准参数内持续运行。质量保证确保体系设计正确，操作正确。质量保证方案包括如上所述的培训、对符合实验室标准的管理审查以及在报告结果之前定期审查结果。有效的质量保证方案对数据的可靠性至关重要。

3.1.11 参与实验室间验证

有可能产生的任何结果都需要与其他来源的结果进行比较。这些结果可能是历史上由其他实验室或其他国家产生的。同样重要的是，从法律的角度来看，表明来自某一特定实验室的数据可以与来自其他实验室的数据相比较。实验室间验证活动是在国际基础上进行的，包括循环测试和实验室间比较等。在这些活动中，同样的样品由多个实验室进行分析，有时使用多种分析方法。比较结果，以确定各实验室之间的平均值和每个实验室与平均值之间的偏差。参与循环测试的实验室间验证可以帮助解决一个不确定的领域，这对于公共卫生监管目的的数据的使用可能是至关重要的。参与循环测试的另一个重要的好处是，如果有高度优先的分析，实验室完全被占用或仪器不能操作,那么其他的实验室可以用来保证数据的可比性。

3.1.12 分析成本

成本似乎是一个主要的考虑因素,但它是决定使用哪个实验室最不重要的因素之一。如果烟草制品制造商已在支付分析费用,这对管理机构不应是一个关键问题,而应作为全面评估的一部分加以注意。

3.1.13 其他客户

在对其他客户(如烟草行业客户)进行分析时,实验室内部可能存在利益冲突。正如第 2.1 节所讨论的,有一些方法可以解决这些问题,但是在做出最终决定时,应该将它们视为一个因素。

3.1.14 位置

实验室与要求分析的国家的距离主要是物流问题,但也可能是进口法的问题。最关键的因素是,是否有可靠的方法将样本从采集地点有效地运送到实验室。由于样品很可能是通过一个普通的载体运送的,这通常不是主要的障碍,但是需要事先评估运输过程和所需的时间。在运输过程中没有被适当存储的样品可能会被损坏,可能会引起关于数据完整性的问题。

一些国家对烟草进口有限制,将此类产品跨越国境运输可能会有问题,但也可以仅允许用于测试目的的产品入境。建议在决定使用另一个国家实验室之前澄清这个问题。还应该考虑到,在附近的实验室可能比在远处的实验室更容易进行检测。

图 2 提供了一种方便的方法来组织实验室评级。加权因素应根据具体国家的需要加以调整。提供了一组可能的权重，但是应该根据需要进行调整。这些权重基实验室在对烟草制品中广泛的待分析物进行精确检测方面尤为成功的因素。分数应由每个实验室决定。然后计算乘积（权重乘以分数）和总和。

实验室名称 _____

因素	权重（1~10）	分数（1~10）	产品（权重×分数）
有烟草分析经验	10		
设备能力	8		
人员资格	8		
能力的广泛性	6		
足够的能力	6		
良好的记录	8		
认证	8		
能够添加新的方法	6		
信息技术系统	6		
质量保证方案	8		
参与实验室间验证	8		
成本分析	2		
其他客户	4		
位置	4		
总和			

图 2　实验室评定表

3.2 WHO 烟草实验室网络

世界卫生组织烟草实验室网络（TobLabNet）是政府、学术机构和独立实验室组成的网络，旨在加强国家和区域检测烟草制品成分和释放物的能力[20]。2005 年 4 月，世界卫生组织无烟草行动组（TFI）根据 WHO FCTC 第 9 条和第 10 条的目标以及 WHO 烟草制品管制研究小组（TobReg）的建议，设立了 TobLabNet。TobLabNet 是各国政府在烟草检测和研究领域提供实验室支持、方法开发和科学信息的主要来源，以满足 WHO FCTC 的要求和需要。

最初，代表 WHO 六大区域的来自 20 个国家的 25 个实验室同意成为 WHO TobLabNet 的一部分。多年来，实验室参与方法验证的情况因国家优先事项和资源的可得性而有所不同。目前参与的实验室名单见表 2。

表 2 烟草实验室网络（TobLabNet）成员实验室名单

WHO 区域	国家	实验室
非洲区域办事处（AFRO）	布基纳法索	国家公共卫生实验室
美洲区域办事处（AMRO）	加拿大	Labstat International ULC
	哥斯达黎加	营养与健康研究和教育研究所（INCIENSA）
	墨西哥	国家公共卫生研究所
	美国	疾病控制和预防中心
		酒类及烟草税及贸易局（TBB）
		伯特利烟草研究公共生卫中心
		弗吉尼亚联邦大学
		国家癌症研究所

续表

WHO 区域	国家	实验室
东南亚区域办事处（SEARO）	印度	卫生服务总局
	印度尼西亚	国家药品和食品控制局
欧洲区域办事处（EURO）	阿尔巴尼亚	公共卫生研究所
	保加利亚	烟草和烟草制品研究所
	芬兰	国家福利和卫生监督管理局
	法国	国家计量和检测实验室
	德国	联邦风险评估协会（BfR）
	希腊	希腊通用化学国家实验室
	爱尔兰	国家实验室
	意大利	欧洲委员会，联合研究中心
	立陶宛	国家公共卫生监测实验室
	荷兰	荷兰国家公共卫生与环境研究所健康保护研究实验室
	俄罗斯	全俄罗斯烟草、马科卡和烟草制品研究所
	西班牙	农业和食品实验室
	瑞士	洛桑大学职业健康学院（IST）
	乌克兰	L. I. Medved 预防毒理学研究中心
地中海东部区域办事处（EMRO）	黎巴嫩	贝鲁特美国大学
	阿拉伯联合酋长国	国家实验室研究中心
西太平洋区域办事处（WPRO）	中国	中国疾病预防控制中心
		烟草安全与控烟技术研究所
	日本	国家公共卫生研究所
	韩国	食品和药品安全部
		韩国疾病控制和预防中心
	新加坡	健康科学机构
	越南	国家职业与环境健康研究所

WHO TobLabNet 的目标是"为管制合规性建设全球烟草制品检测和研究能力，研究和制定成分和释放物检测的统一标准，分享烟草研究和检测标准及结果，告知与烟草制品使用有关的风险评估，并协调报告这些结果，使数据可以转化为有意义的趋势信息，可以

在不同国家和不同时间进行比较"[21]。

为了实现这一目标,各实验室在不同领导实验室的指导下,协同合作,相互支持。WHO TobLabNet 积极工作,向寻求发展和改进烟草检测实验室的国家政府提供咨询,作为提高能力和确保一致性的手段。

WHO TobLabNet 通过 WHO FCTC 秘书处,在 WHO 的主持下进行工作,以实现 WHO FCTC 规定的目标。最近,这项工作涉及对商业卷烟和其他烟草制品的高优先级成分和释放物进行检测的方法开发和验证。此外,各实验室对方法进行了循环测试,用于测量这些方法的实验室间可重复性。目前的成分清单及其状况见表3。

表3 当前 TobLabNet 方法开发现状

成分	待分析物	基质	分析方法	现状
烟碱	烟碱	烟草	GC/FID[a]	已验证
氨	氨	烟草	离子色谱/电导检测	已验证
保润剂	丙二醇 甘油 三甘醇	烟草	GC/FID(GC/MS)[b]	已验证
TNCO	焦油、烟碱和一氧化碳	烟气	GC/FID 烟碱 GC/TCD[c](焦油) 非色散红外分析(CO)	已验证
TSNAs	NNN,NNK,NAT,NAB	烟气	HPLC/MS-MS[d]	已验证
BaP	BaP	烟气	GC/MS	已验证
VOCs	苯 1,3-丁二烯	烟气	GC/MS	已验证
金属羰基化合物	甲醛 乙醛	烟气	HPLC/DAD[e]	已验证

a. 气相色谱/火焰离子化检测
b. 气相色谱/质谱
c. 气相色谱/热导检测器
d. 高效液相色谱/串联质谱
e. 高效液相色谱/二极管阵列检测器

寻求关于建立烟草制品检测机制信息的各国政府，建议与世界卫生组织联系，寻求建议和指导。根据资源的可得性，WHO TobLabNet 可能会提供培训和能力建设支持，以便开始烟草检测或拓展现有能力，包括现有成员，以及未来成为成员的实验室。

3.3 协议和法律与伦理问题

如上所述，国家监管机构在计划与外部实验室签订协议时应考虑其具体的法律和伦理问题。在签署协议和进行分析之前，应该了解和解决这些问题。非常重要的是，监管机构获得的数据必须足以满足其目的；所有与数据质量有关的使用数据的要求应在其他协议中清楚地描述。

与外部实验室达成协议时，首先要考虑的是确保实验室独立于烟草行业。WHO FCTC 第 5.3 条敦促缔约方保护烟草控制政策不受"烟草行业的商业和其他既得利益"的影响。WHO TobLabNet 采用严格的成员资格标准，不包括完全或部分由烟草公司拥有的实验室，也不包括由烟草行业雇用或隶属于烟草行业的高级管理职位人员担任的实验室。从烟草行业获得资金的实验室还必须表现出独立性。这些利益冲突要求确保了检测政策的公共卫生目标不会受到损害。

另一项考虑是法律要求将测试数据作为法律程序的证据，或作为监管行动的基础。例如，应该从一开始就了解到，实验室具有必要的认证和数据质量的证据，以便在采取行动时使用这些数据。监管链可能是确保数据是来自准备测试材料分析的关键因素。样本完整性的保证是法律上接受数据的关键因素。

某些实验室可能考虑与专有分析相关的背景信息。对实验室方法、质量控制结果、方法验证的发现或关于循环测试结果信息的详细描述可以视为私有的,而不公布。如果需要这些信息作为监管证据的一部分,那么从一开始就应该清楚,实验室必须提供这些信息。否则,在分析完成后公布,将极大地限制数据的有效使用。

另一个签约前的考虑是报告分析结果。实验室测量通常涉及对每个样本进行多重重复分析。换句话说,为了提高准确性,考虑到由烟草等产品制成的商业产品的可变性,并评估可重复性,进行多项测量,并平均得出最终结果。通常情况下,这可能包括3~20个副本,以获得单个最终结果。要进行的重复分析的数量是样品可用性、成本(重复越多成本越高)和数据质量(更多的重复减少了随机变化)之间的平衡[4]。监管机构应该考虑如何最好地平衡数据使用的需求和分析的成本。在选择实验室时,通常进行的重复数量应该是整体考虑的一部分。

在提供给客户的报告中,实验室可以报告所有的副本,也可以报告最终平均值。监管机构应要求报告所有重复样品,以便充分评估数据质量,并根据具体应用情况进行适当的统计分析。这应包括在与实验室的协议中,以便保留和报告所需的资料。

在使用外部实验室时,重要的是要了解这些数据是否以及何时对请求和资助分析的单位以外的人员可用。实验室应有足够的防火墙,以便与包括烟草制品制造商在内的其他相关方达成的商业协议不会影响监管机构的分析。在同意使用实验室之前,监管机构应该

---

[4] 关于这个问题的更多细节请见附录。

调查实验室是否存在潜在的利益冲突。此外，监管机构应注意公共获取或信息自由法，这可能影响有关各方对数据的访问。这些法律可能不会影响监管机构是否使用特定的外部实验室进行检测，但它们应该意识到可能出现的问题。

即使了解到这些分析将由行业提供资金，报告结果也至少有三种选择：只报告政府，政府和行业同时报告，或者只报告行业，然后向政府提供数据。从可靠性的角度来看，如果实验室的结果符合法律或法规中规定的要求，最好直接向政府报告。这可能是通过向监管机构报告结果，或同时向监管机构和产品被评估的制造商提供。直接向政府机构报告有助于向实验室强调。监管机构是结果的最终使用者，是在何处进行分析的决策者。由于缺乏信任，同时报告是所有有关各方最可能接受的选择。但应避免实验室向制造商报告，制造商随后向监管机构提供这些数据，因为这为数据操纵提供了机会，并产生了不必要的不确定性。

3.4　样本量估计

在选择实验室时，实验室能力是一个重要的考虑因素，因为当时机成熟时，及时的结果对于向决策者提供数据至关重要。但是过剩的实验室能力是有代价的。一个实验室要为大量的样品负荷或额外的要求做好准备，必须具备多余的设备和人员。即使在不使用时，也必须支持这些资源，因此管理费用会增加。非常重要的是，监管机构要对预期的样本负荷提供良好的估计，以便实验室能够提前准备,及时做出反应。对样品负荷或样品交付时间的估计不佳导致结果报告或资源浪费,费用必然转嫁给客户。

为了最好地优化可用资源,抽样和分析请求应该在全年中进行,而不是每年一次。有几种有效的方法可以做到这一点。制造商可能被要求在一年中不同的时间提交样品,可能会将这些分散在四个季度。或者制造商可以在每年的每个季度提交四分之一的品牌/子品牌进行分析。这将分散实验室的工作量,允许更有效地使用资源,并有助于控制每个分析的成本。

在选择实验室时,重要的是要基于能力和满足规定时间范围的能力进行评估。必须对不可预见的情况进行一些考虑,在估计样本报告预期时应该考虑到这一点。监管机构需要考虑提交后多久才能取得结果,以及是否需要提出特别优先要求。在与实验室签订合同时,应将其包含在协议中。

3.5 成 本

与其他监管活动相比,烟草制品的成分和释放物分析的成本较高。无论制造商是直接向实验室提交样品,还是先将样品提交给监管机构,再提交给实验室,所有费用都必须由烟草制品制造商承担。如果制造商认为检测的成本过高,他们可以选择不推广自己的产品或减少在特定市场上销售品牌/子品牌的数量。任何一种情况下,在一个国家销售的品牌/子品牌数量的减少都可能导致烟草制品的总体使用减少和公共卫生的改善。

根据所进行的分析基质（烟草或烟气）、使用的设备以及重复的数量，每个单独的品牌／子品牌的每个分析测试的成本可以从数百美元到数千美元不等。烟气的测量通常比烟草的测量更昂贵，因为用适当的方法产生和收集烟气的过程为分析过程增加了一个额外的步骤。有些分析可以用各种分析设备进行。例如，分析主流烟气中的苯并[a]芘可以使用红外光谱、气相色谱与质谱联用、高效液相色谱与荧光检测联用、高效液相色谱质谱联用或高效液相色谱与串联质谱联用。当实验室可以执行足够的交叉验证时，使用其他分析仪器进行验证成为可能。购买和维护这些设备的成本不同，因此分析的成本也不同。使用更精密和灵敏的分析设备可能有一些关键的好处（灵敏度和选择性），但是在不需要的时候应该避免使用成本更高的设备。在同意进行更高的成本分析之前，应该对分析测量结果的适用性进行评估。

在某些情况下，可以用同样的分析方法同时测量多个成分。例如，大多数用于TSNAs的分析方法都使用相同的方法来测量这四种化合物。如果只测量其中一个TSNAs，成本不会显著降低。有些多环芳烃、重金属、醛和许多挥发性有机化合物也是如此。因此，测试和报告的总成本不那么依赖于所报告的成分数量，而是依赖于必须执行的分析方法的数量来进行测量。

3.6 案例研究：加拿大

加拿大是一个很好的例子，它成功地将外部实验室用于研究目的和方法开发来支持烟草控制。20世纪70年代末，加拿大卫生部开始承包一家独立实验室⑤（不隶属于烟草行业）的服务，以测试加拿大市场上各种烟草制品的成分和释放物。在20世纪80年代，确定在烟草制品成分和释放物检测中需要测量更大量的待分析物后，加拿大卫生部要求建立同样的独立实验室开发实验方法，这项工作一直持续到20世纪90年代，并提出了涵盖烟草制品成分中20种待分析物和释放物中40种待分析物的方法[22]。

2000年开始实施的《烟草报告条例》将这些纳入"官方方法"，制造商必须将其用于向加拿大卫生部报告[23]。根据《烟草报告条例》第4条[22]，烟草制品制造商必须使用经国际标准化组织标准ISO/IEC 17025认证的实验室报告结果，该实验室对检测和校准实验室的能力提出了一般要求。通常，独立的实验室被制造商保留以履行其报告义务。加拿大卫生部门继续根据需要使用独立的实验室来支持分析发展项目。

使用单一的分析实验室和一套确定的分析方法具有明显的优势。当同一实验室使用相同的方法进行测量时，不确定性显著降低，并且对所有结果的直接可比性有了更大的信心。这种方法的缺点是依赖于单个实验室。例如，如果该实验室停止操作，将必要的分析转移到另一个实验室将面临挑战。此外，如果只使用一个实验室，可能会对样本分析能力产生担忧。如果只确定一个实验室，那么最好让实验室提供一份备份计划，说明可以采取哪些步骤来减少任何能力损失，并确保操作的连续性。

---

⑤ 加拿大安大略省Ontario的Labstat (http://www.labstat.com/servicesoverview.html)。

3.7 循序渐进的过程

1. 做自己的内部研究。

（1）为烟草检测确定一套合理的第一套分析方案，以解决国家的优先事项。

（2）确定数据预期使用的分析需求。

（3）确定估计的初始工作负荷（在什么时期内的样品数量）。这可能取决于正在销售的品牌/子品牌的数量以及检测的频率和时间表。

2. 与另一个有烟草制品检测经验的国家的监管机构讨论方法。

3. 评估可用的实验室。

（1）检查上面列出的要求并对可用的实验室进行排名。

（2）评价实验室满足分析准确度、可重复性、灵敏度和选择性要求的能力，以确定满意的实验室。

（3）使用能力需求估计来确定可接受的实验室。

4. 与确定的实验室讨论预期样品工作量、周转时间、报告要求等方面的特殊要求。

5. 完成任何必要的合同协议。

6. 如果合适的话，与制造商沟通：

（1）哪些要测试，何时需要测试？

（2）哪些实验室是可以接受的？

第4章　利用现有的内部检测实验室

在某些情况下，国家政府可能已经有实验室来检测非烟草消费品。可能是测试药物、化妆品或进口商品的纯度或验证治疗药物的含量水平，也可能有实验室测试空气或水等环境样本。其他消费品或环境样本的政府检测实验室的存在为随时发展政府烟草检测能力提供了一个机会。

烟草监管机构可能会发现，已经在经营烟草检测实验室的政府组织愿意合作扩大他们的能力，并将一些烟草设计、成分和释放物测试作为其分析组合的一部分。因为他们可能已经超负荷工作，资源不足，这可能需要一些细致的说服，强调这可能会给他们的活动带来好处。解释扩大检测范围至烟草领域将如何有利于他们目前的计划，可以帮助说服他们。

如果可能的话，与现有的政府测试实验室合作有几个好处。现有的实验室可能有设备、消耗品、接受过培训的人员、质量保证和信息系统。在建立实验室时，单独使用设备是昂贵的。分析设备的价格可能高达每件产品数十万美元，维护和维修是一个持续的需求。如果目前的实验室已经有设备，并已为维护和维修编列预算，这些费用已包括在内，对烟草制品的进一步分析不会大幅增加费用。

实验室人员是关键的资产。要培养一名有效率的分析人员，需要多年的培训和经验。虽然可以直接雇佣有经验的分析人员，但由于资源有限，可能是一个重大挑战。如果一个实验室已经有经验丰富的分析人员，就像现有的测试实验室一样，培训他们进行烟草制

烟草制品管制
实验室检测能力建设

品分析方法所需的时间将大大减少。如果优先考虑快速获得结果来说明烟草制品测试的价值，那么与现有实验室合作比从头开始创建实验室要好得多。经验丰富的分析人员的价值怎么强调都不为过。进行分析的工作人员的知识和经验是决定是否能够有效地获得可靠结果的关键因素。

同样，如果需要，使用现有的实验室可以允许测试项目以有限制的方式启动。在从头开始建造实验室时，必须考虑到实验室的规模和未来多年的预计资源。因此，最初的设计必须考虑到预期的能力和容量的增加。不可避免的是，这会导致在项目开始时过度建设造成大量不用的配置。对于一个刚刚开始的项目来说，这可能是一个挑战。通过使用现有的实验室设施，烟草测试可以在有限的水平开始，然后随着项目的扩展以控制的方式增长。

这种方法的另一个重要优势是，它可以为现有的实验室提供额外的资金来源。实验室总是必须平衡他们的仪器容量和工作人员的数量和预期工作量和资金。样品负荷的周期性变化一直是实验室管理的挑战。通过扩大流向实验室的资金来源和数量，可以更好地平衡样本负荷的变异性。这是与当前实验室管理讨论此方法时可以展示的一个重要优势。虽然资源需求将有所增加，但额外的资金来源可能有助于解决样品分析需求和资金的变化。

除了使实验室的资金来源趋于平稳之外，额外的资金还可以支持购买更多的设备、雇用新的工作人员和扩大其他本来不太可能实现的能力。大多数用于烟草制品成分和释放物分析的设备可用于其他工作。因此，如果允许的话，烟草资助可以用来升级分析仪器，也可以用来分析环境样本或药品和其他消费品。这将为目前的实验室提供一个重要的优势。就协同作用的可能性而言，这两个项目都

第4章 利用现有的内部检测实验室

将受益。

这种方法有一些缺点。由于目前的方案责任，该方案很可能将烟草分析视为较低的优先事项，至少在一开始是这样。必须讨论优先事项并达成一项协定，澄清如何解决分歧以及两个方案如何共同合作以满足所有需要。否则，这可能是一个重大问题。

此外，对烟草制品的检测将有一些在其他检测方案中使用的分析中没有的要求。最明显的例子是需要控制温度和湿度的设施，这些设施需要测试传统烟草制品。现有的实验室管理人员可能不愿意去实现这些新功能。应该事先讨论，并确定一个预期的时间表和计划，防止将来产生误解。

4.1 要求（实验室设备、人员、总成本）——如何确定合适的实验室

需求直接依赖于确定为最高优先级的初始分析。如果这些分析不能用现有设备进行，那么使用现有的检测实验室的好处是有限的。但是前面列出的大部分分析设备（见表3）应该可以在典型的分析实验室中使用。

在评估可能适合增加烟草分析能力的现有实验室时，第一步是对现有的分析设备进行比较分析。表4列出了在检测实验室中常见的用于分析烟草制品设计、成分和释放物的设备清单。这个表比表3中列出的要广泛，并且这个表中有一些重复，因为一些分析物可以由多个分析仪器测量。

在此之前，世界卫生组织确定了用于发展烟草检测实验室的有用设备[24]。该清单可用于评估现有政府实验室的仪器设备，以评估根据国家的优先次序，哪些现有实验室可能配备最好的设备，以及哪些设备可能仍需为特定的烟草测试应用而采购。

为了进行烟草制品特异性分析，可能需要购买一些额外的设备。该设备应在实验室发展的总体计划中加以考虑。

- 吸烟机（一氧化碳非色散红外分析仪），约 20 万美元。
- 环境室，约 3 万美元。

第 3.1 节描述了选择外部检测实验室的标准。除了在烟草制品检测方面的经验外，这些标准也适用于评估测试其他受管制材料的内部实验室。评级电子表格（图 2）也可以用来识别在这种情况下需要考虑的方面。特别是，除了可用的设备之外，在评估现有检测实验室的适用性时，还应评估以下内容：

- 足够的工作台空间。
- 有效的信息技术系统，可用于跟踪样本、限制数据转录、有效报告和结果存档。
- 符合认证要求的质量保证计划。
- 符合当前和预期仪器要求的环境控制（温度和湿度）。
- 足够的电气系统，满足仪表制造商的预期要求，确保仪表正常运行。
- 训练有素、经验丰富的员工，他们有编制可靠的分析测试结果的历史记录。

第 4 章 利用现有的内部检测实验室

表 4 烟草试验室的分析设备

仪器	目的	近似成本(美元)
冰箱	存储样品	1 000
分析天平	称量样品,"焦油"	10 000
压降装置	卷烟通风	40 000
离子色谱/电导检测	烟草中的氨	50 000
连续流动比色分析	烟气中的氰化氢	60 000
化学发光氮氧化物分析仪	氮氧化合物	50 000
气相色谱/火焰离子化检测	烟草/烟气中的烟碱	100 000
气相色谱/热能分析	烟草/烟气中的TSNAs	150 000
高效液相色谱/紫外检测	烟气中的羰基化合物	100 000
高效液相色谱/荧光	BaP,烟气中的酚类化合物	100 000
气相色谱/质谱	烟气中的烟碱、VOCs、羰基化合物、多环芳烃、香味物质、芳香胺	150 000
高效液相色谱/串联质谱	烟草/烟气中的TSNAs	250 000
原子吸收光谱	烟草/烟气中的金属	50 000
电感耦合等离子体-原子发射光谱	烟草/烟气中的金属	70 000
电感耦合氩等离子体质谱	烟草/烟气中的金属	200 000

注:这些费用仅为近似值,并可能因国家的具体差异而有很大差异
挥发物:苯、1,3-丁二烯、丙烯腈
羰基化合物:丙烯醛、甲醛、乙醛、巴豆醛
金属:砷、镉、铬、铅、汞、镍、硒

4.2 认 证

所有的分析实验室都应该得到国际或国家机构的认证。实验室的标准是 ISO/IEC 17025[25]。这适用于所有形式的检测实验室,包括

药物实验室、环境实验室、烟草制品实验室和其他实验室。

ISO 17025 规范一般实验室能力和管理。它评估实验室是否有记录方法、人员资格和培训、测量验证和最小化误差的体系和协议。它足够广泛，允许实验室使用标准的方法、广泛接受的方法和实验室开发的方法。

ISO 17025 不是设计用来评估实验室使用的方法是否准确、可重复和灵敏以使测量适合于特定的应用。例如，它不评价哪一种分析方法最适合某一特定分析。这通常是通过实验室内和实验室间的验证来实现的。因此，认证是获得实验室能力的必要因素，但不是充分因素。

由于 ISO 标准的一般性质，它们被认为是实验室的最低要求，但还不足以证明实验室能够提供准确和可重复的分析结果。只有通过第 3.1 节所述的完整质量保证方案才能证明这一点。

4.3 案例研究：新加坡

新加坡的卷烟检测实验室（CTL）就是一个很好的例子，它利用了现有的政府实验室功能。CTL 与制药实验室和化妆品实验室组成了新加坡卫生科学管理局的制药部门。CTL 成立于 20 世纪 80 年代末，其任务是检测主流卷烟烟气中的焦油和烟碱，以支持烟草法规的遵从。它后来利用药品和化妆品实验室现有的分析设备扩大了处理焦油和烟碱以外的有害物质的范围。这种方法使实验室能够以最低的额外成本扩大其能力。

该实验室除了作为 WHO TobLabNet 的一部分进行培训来协助

其他国家的能力建设外,还支持需要检测设施的国家和地区的烟草测试倡议,来支持他们的烟草管制框架,这些国家和地区包括斐济、文莱、汤加、所罗门群岛和萨摩亚。这项工作利用现有的检测实验室设施来建立能力并支持烟草法规的遵守,为其他国家提供了良好的模式。

4.4 循序渐进的过程

1. 做自己的内部研究。
(1)为烟草测试确定合理的第一套分析方案,以解决国家的优先事项。
(2)确定对数据的预期使用的分析要求。
(3)确定估计的初始工作量(在什么时期内的样本数量)。这可能取决于正在销售的品牌/子品牌的数量以及测试的频率和时间表。
(4)确定进行分析所需的仪器(见表4)。

2. 与另一个有烟草制品检测经验的国家的监管机构讨论方法。

3. 参观其他已经在进行消费品测试的政府实验室。
(1)检查上述要求并对现有实验室进行排名。
(2)评价实验室满足分析准确度、可重复性、灵敏度和选择性要求的能力,以确定满意的实验室。
(3)利用能力需求估计数来确定可接受的实验室。

4. 酌情与其他政府组织协商,以取得在测试方面的合作协议。

5. 与确定的实验室讨论预期样本工作量、周转时间、报告需求、

优先级冲突等方面的特殊需求。

    6. 完成任何必要的合同协议。

    7. 公司沟通：

      （1）哪些要测试，何时需要测试？

      （2）哪些实验室是可以接受的？

第 5 章　开发烟草专用检测实验室

在以下讨论中，独立烟草测试实验室是指在政府系统内的实验室，该实验室不与其他项目共享资源（设备和人员），尽管它可能被安置在同一个物理设施中。开发一个独立的政府烟草测试实验室可能是一项重大挑战，除非从现有的实验室能力开始，因为所需的时间和资金可能相当可观。维护管理支持，直到项目完成也是一个挑战。一些国家试图在不扩大现有能力的情况下建立独立的实验室设施，但迄今为止这些设施都没有成功。能够建立独立的烟草测试实验室的组织通常在另一个实验室测试方案的基础上建立这些能力，使其能够自我维持和独立（见第 5.4 节的案例）。

5.1　要求（基础设施、实验室设备、人员、总成本）

所需的设施占地面积、设备和资源，取决于测试方案的预期范围。在过程的早期就这些要求做出明确的、深思熟虑的战略决定是关键的一步，并将极大地影响整个方案的长期成功与否。这无论如何强调都不为过。

希望建立具有广泛能力的实验室的组织必须预见预期的空间需求。在此之前，TobReg 曾在 2004 年为一家烟草检测实验室提供建议[1]。本书对检测实验室提出如表 5 所示建议。

表 5　实验室的空间要求

区域类型／工作场所	最小表面积 (m^2)	扩展实验室表面积 [a] (m^2)	条件
制备实验室	20	60	给水和排水要求 金属分析将需要一个单独的"洁净室"
烟气实验室	20	60	包含吸烟机 温度和湿度调节（22℃ ±2℃和60%±5%）
仪器室	30	80	空调；专用仪器将需要额外的通风和其他特定的环境控制
办公室	20	40	
存储室	15	25	
共同区域	15	25	
杂物间	15	40	
总计	135	330	

a. 扩展的实验室将包括执行所有建议的化学成分分析的必要设备

这个空间需求描述仅仅是基于测试需求的预期。一个不打算进行许多分析的方案则需要较少的空间，一个打算进行更多分析的方案将需要更多的空间。强烈建议，在方案做出最后空间决定之前，应先访问一个目前正在运作的烟草检测实验室，以便更好地了解预期的要求。

表 4 所列清单列出了可能需要的基本设备，以提供实验室的分析能力，以便确定为高度优先的分析。扩充后的实验室可能需要第 4.1 节所述的其他设备。除了表 4 所列的设备外，可能还需要标准的实验室设备。

一个合格的、受过良好培训的工作人员是烟草制品测试成功所必需的。对许多分析来说，专业培训是必要的。就楼层面积和设备而言，工作人员的数目将取决于分析方法的预期数目以及预计在一年期间进行分析的样本数目。在上文提到的同一份文件[1]中，TobReg还根据每年对150个品牌/子品牌进行检测的典型实验室提供了人员配置的初步建议。TobReg建议如下：

- 1名烟气实验室经理；
- 2~3名烟气技术员，熟悉吸烟机的操作和维护；
- 2~3名分析化学家，应该具备广泛的仪器知识；
- 1名质量控制经理，负责对数据和方法进行监督和控制，并精通统计和数据报告。

如果期望增加抽样工作量，则应按比例增加人员。这是最低限度，不包括为实验室工作提供支持的行政工作人员或其他非技术人员。

5.2 信息技术系统

一个高效的IT系统对及时地减少错误并报告结果至关重要。IT系统的重要性常常被那些不熟悉测试实验室需求的人所忽视。与现有测试实验室进行背景讨论的一个重要部分应该包括讨论当前IT系统的功能和需求。一个实验室的IT系统至少应该包括以下能力：

- 允许登录新样品。
- 能够根据变化的优先级对样品进行排序。
- 通过分析过程跟踪样本。
- 允许在适当的情况下进行自动数据计算（大多数分析设备都有内部系统来调度、处理和帮助分析人员分析原始数据，但这些系统必须与整个IT系统兼容）。
- 独立于分析人员评估质量控制结果。

- 重新安排不符合质量控制要求的样品分析。
- 报告最终结果。
- 归档所有数据。
- 备份所有数据。

5.3　数 据 验 证

在报告任何数据之前,必须对系统和流程进行仔细的数据验证。对于已经有经验的、与其他管制方案有关的检测的外部实验室,应该已经有了体系。对于一个新建立的独立的内部实验室,需要开发体系。

5.3.1　质量控制材料分析

质量控制材料是在每次分析运行中引入的样品,以确保系统正常运行。对质量控制材料进行评估有完善的原则[26]。当从质量控制材料的分析中得出的结果偏离统计上可接受的范围时,结果被拒绝,必须对分析系统进行调查。

5.3.2　准确性和可重复性的系统检查

在使用任何方法进行附录所述的分析之前,应确定分析方法的准确性和可重复性。但设备或其他条件的改变,可能导致这些最初的结果不再准确。通过分析已知的参考材料,通过重复分析,对准确性进行定期检查,以确认系统在原始条件下继续运行,初始措施

第 5 章　开发烟草专用检测实验室

是有效的。

5.3.3　长期趋势分析

由于这些趋势的性质，分析结果的长期偏差很难识别。其他旨在在较短时间内监测的质量保证系统可能不会识别长期偏差。系统应该到位，以确保数据的长期偏移被识别和修正。

5.4　案例分析：CDC

1994 年，美国疾病控制与预防中心（CDC）的吸烟与健康办公室（OSH）与国家环境健康中心（NCEH）实验室的工作人员接触，为满足烟草行业向 OSH 提供的信息审查的某些监管要求提供支持。因此，确立了明确的测试任务。当时，该实验室已经有一名具有 10 年经验的专门工作人员，使用先进的分析仪器用于生物监测，评估烟草制品使用者和非使用者的接触情况。实验大楼已经到位，有必要的环境控制和不间断供电。此外，设备的服务合同已经签订，现场有替换零部件，以便迅速进行维修。实验室已经有了广泛的质量控制方案、统计和 IT 支持。最后，有一个强有力的支持结构，它不依赖于快速的结果，而是了解战略方法的必要性。

NCEH 实验室的几名工作人员参观了位于加拿大基切纳的 Labstat 公司的私营烟草分析实验室。工作人员亲切地解释了一个成功的烟草测试实验室所需的所有要求（环境控制、设备、员工等）。这使 NCEH 实验室的工作人员了解到，要成功地发展实验室能力以测试烟草制品，还需要什么其他要求。该实验室购买了一个环境室、一个吸烟机和专门用于烟草分析的烟草制品设计实验设备。这种特殊设备补充了进行分析测量所需的更通用的实验室设备。在很短的时间内，NCEH 烟草实验室获得了与 NCEH 实验室其余部分分离的设备、实验室空间和资源，使设备不再在方案之间共享，而是专门用于烟草制品测试和研究。

在接下来的几年中发生了几次具体的关键事件，包括 1995 年菲利普·莫里斯召回事件[27]，为 NCEH 实验室证明疾病控制中心烟草控制方案的价值提供了机会。从那时起，该实验室的规模不断扩大，因此具有广泛的能力，并为美国食品和药品监督管理局（FDA）的 CDC 任务和烟草制品研究提供分析支持。该实验室还作为一个培训实验室，与 FDA 自己的东南地区实验室合作，开发和验证合规检测新方法。

CDC 烟草实验室是从已经存在的能力中衍生出来的。但实验室不再与其他项目共享设备和人员。这一过程持续了数年，使实验室得以在最初的投资中立足。它还允许实验室随着需求和资金的增加而成长。

5.5　循序渐进的过程

1. 做内部调查。

（1）确定一套合理的第一批烟草检测分析，以解决国家的优先事项。

（2）确定对数据的预期使用的分析要求。

（3）确定估计的初始工作量（在什么时期内的样本数量）。这可能取决于正在销售的品牌/子品牌的数量以及检测的频率和时间表。

（4）确定进行分析所需的仪器（见表4）。

2. 参观已经在做烟草制品测试的其他实验室，以更好地了解烟草制品测试的要求。

确定空间、设备和人力资源需求。确定需要的专业设备和培训。

3. 确保建立和维持实验室能力的长期支持和资金的行政保证。

4. 与建立的烟草检测实验室密切协商。

（1）根据需要发展实验室设施。

（2）雇佣有经验的员工。

（3）购买分析设备。

5. 派遣员工到已建立的实验室接受培训。

6. 实施上述有限数量的实验室方法。

7. 通过评估分析的准确度、可重复性、灵敏度和选择性，在实验室内对方法进行验证。

8. 参与实验室间验证活动或与有经验的实验室交换样品进行分析。

9. 通过重复上面的 5~8 来扩展功能。

10. 公司沟通：

（1）哪些要测试，何时需要测试？

（2）哪些实验室是可以接受的？

第6章 资源：WHO TobLabNet 成员（标准、优势和程序）

WHO TobLabNet 实验室可以作为任何考虑通过上述机制开发测试能力的监管机构的重要资源。在考虑如何将实验室纳入烟草管制计划时，访问 WHO TobLabNet 实验室可能是一个宝贵的机会，看看其他国家是如何应对这一挑战的。

世界卫生组织成员实验室可以提供关于空间需求、环境（电力需求、空调、水质等）需求、分析仪器和人员需求的非常重要的信息。通过访问 WHO TobLabNet 成员实验室，政府监管机构可以直接了解如何设计实验室，并与 WHO TobLabNet 成员讨论他们的经验教训，以及通向成功的步骤。

通常，购买先进的分析设备包括制造商／供应国／供应商的一些培训。这种培训通常包括仪器操作和维护的基础知识和软件的使用。但是，这种培训不够具体，不能使工作人员对烟草制品的设计、成分和释放物进行测量。在时间允许的情况下，WHO TobLabNet 成员可以为分析人员提供培训，培训内容是如何制定针对烟草制品的具体测量，以补充仪器制造商／供应国／供应商的培训。

最有效的是，培训应该在设备购买和安装后进行，并且在分析人员有一些实际操作经验后。如果分析人员在设备安装之前接受培训，那么培训经验就不会那么有效，因为他们无法根据操作设备的经验提出更实际的问题。培训可在发展中或新建立的实验室，或在 WHO TobLabNet 实验室进行。

第 6 章 资源：WHO TobLabNet 成员（标准、优势和程序）

在发展中/新建立的实验室中进行培训的好处是，培训人员以后要使用的任何系统都有。但为了有效利用培训时间，必须安装和操作分析设备。参加培训的员工应该是实际操作设备的分析人员。没有执行分析设备日常操作的经理或官员不适合参加培训。由于信息的技术性质，他们将无法有效地交流经验。

可以向 WHO TFI 的工作人员提出培训要求，他们可以建议适当的实验室。WHO TFI 还正在开发一系列在线培训模块，以利用世界卫生组织现有的 TobLabNet 标准操作规程，在 ISO 和严格吸烟条件下测量烟丝和卷烟主流烟气中的优先有害物质。一旦培训平台启动，WHO TFI 网站将提供更多信息。

作为一个国际实验室网络，世界卫生组织可以作为一个有价值的活动来源，帮助发展和展示成员实验室的能力，以进行有效的测量。这方面的一个例子是由两个或两个以上独立的实验室/两个独立的分析方法测试有限数量的样品材料，例如，如果一个实验室正在操作一种已知的方法或开发一种新方法，在多个实验室对共享样本进行测量可以帮助证明报告结果的准确性。

此外，实验室间活动可以成为参与实验室的宝贵验证来源。这包括周期性的循环测试，由世界卫生组织烟草控制框架公约（FCTC）指导，或由 WHO TobLabNet 作为工作产品开发。之前的循环测试已经形成了可在 TFI 网站上获得的已发表的文件[28]。

WHO TobLabNet 定期召开网络成员会议，以鼓励信息交流和规划未来的联合项目。这对新实验室来说是很有价值的，可以作为解决问题和比较经验的论坛。世界卫生组织在其网站[28]上列出了 WHO 实验室成员资格的标准，并列示如下：

- 机构在国家卫生、科学或教育结构中的地位；

- 与在该国家或地理区域内活动的烟草控制社区合作的证据;
- 不受与测量结果中有重大的财务利益的组织或实体的关系的过度影响;
- 其科学和技术领导的素质,以及其员工的数量和资格;
- 机构有能力、容量和准备程度,能够单独或在实验室网络内为TobLabNet方案活动作出贡献;
- 有烟草产品测试、研究经验或明显意图,获得烟草制品检测或研究能力,例如承诺培训人员和升级设备;
- 机构在人员、活动和资金方面的预期稳定性;
- 机构及其活动与TobLabNet方案优先事项的技术和地理相关性;
- 机构与该国其他机构以及在国家、区域和全球各级工作关系;
- 有关机构在国家和国际层面的科学和技术地位。

此外,为了防止利益冲突,以下列出了其他成员标准[29,30]:

- 实验室不应全部或部分归烟草公司所有,但允许由拥有或经营国家烟草行业的国家政府拥有或经营的实验室。
- 从烟草行业以提供服务的形式获得融资的实验室必须证明其独立于烟草行业。对于这些组织,需要利益冲突表格。
- 如果是上市公司,烟草行业占总股份的比例不应超过10%。
- 实验室不应有烟草公司聘用的任何董事会成员或高级管理职位的人员,其中包括咨询职位等。还包括向烟草公司提供无报酬的咨询或建议,这些咨询或建议可能因履行了未来工作的承诺而引发冲突。
- 实验室可以有烟草公司作为客户,但不能是唯一客户。

总　　结

烟草制品的检测能力对于那些试图通过控制烟草制品来减少烟草使用造成的死亡和疾病的国家来说是一个有价值的工具。对世界卫生组织《烟草控制框架公约》缔约方来说，烟草制品成分和释放物的监管（第9条）和烟草物品披露的监管（第10条）是可能有助于减少烟草制品需求的关键措施之一。虽然它本身不是一个答案，但它可以用于通知和建立其他监管活动，如产品审查、产品标准、包装和标签法规、公共教育或作为信息通知立法决策者。

开发测试能力的方法应该根据明确的目标进行战略性的考虑。这些目标因国家而异，只能通过考虑个别的国家目标来确定。通过有效地利用有限的资源来实现最大的效益，这种初步的基础工作将获得多次回报。

虽然从头开始建立一个实验室是可能的，但尝试过这种策略的国家并没有成功。建议各国要么与已经在测试烟草制品的实验室签订合同，要么从具有测试其他消费品（如药品）或环境样品经验的现有实验室建立能力。这一方法已在若干国家获得成功，并为实现一个国家的目标提供了最佳机会。

世界卫生组织的工作是为了支持现有能力，并协助发展新的国家烟草制品检测能力。WHO TobLabNet 可以以多种方式帮助实验室发展，有兴趣发展新实验室的国家应与 WHO TFI 联系，WHO TFI 可使它们与 WHO TobLabNet 实验室联系，并协调活动，以支持制定检测方案。这种接触在开发实验室的过程中应尽早进行。

参 考 文 献

[1] Guiding principles for the development of tobacco product research and testing capacity and proposed protocols for the initiation of tobacco product testing. Geneva, World Health Organization.2004(http://www.who.int/tobacco/publications/prod_regulation/goa_principles/en/).

[2] Brazil-Flavoured cigarettes banned. Geneva, WHO FCTC, March 2012 (http://www.who.int/fctc/implementation/news/news_brazil/en/).

[3] Canada (Legislative Summary of Bill C-32: An Act to amend the Tobacc Act), Publication Number LS-648E, 2009 (https://lop.parl.ca/About/Parliament/Legislative-Summaries/bills_ls.asp?ls=C32&Parl=40&Ses=2).

[4] Canada (Order Amending the Schedule to the Tobacco Act (Menthol) P.C. 2017-256 March 24, 2017). (http://gazette. gc.ca/rp-pr/p2/2017/2017-04-05/html/sor-dors45-eng. php).

[5] Regulation, the Tobacco Atlas. (http://www.tobaccoatlas.org/topic/regulations/, accessed 15 January 2018).

[6] Brewer NT, Morgan JC, Baig SA, Mendel JR, Boynton MH, Pepper JK, et al. Public understanding of cigarette smoke constituents: three US surveys. Tob Control. 2016 Sep;26(5):592-599.

[7] Morgan JC, Byron MJ, Baig SA, Stepanov I and Brewer NT. How

people think about the chemicals in cigarette smoke: a systematic review. J Behav Med. 2017 Aug;40(4):553-564).

[8] Canada (Regulatory Impact Analysis Statement, Canada Gazette, Part I: Vol. 145 2011). http://www.gazette.gc.ca/rp-pr/p1/2011/2011-02-19/html/reg3-eng.html

[9] Kelley DE, Boynton MH, Noar SM, Morgan JC, Mendel JR, Ribisl KM et al. Effective Message Elements for Disclosures about Chemicals in Cigarette Smoke. Nicotine Tob Res. 2017 May 17 doi: 10.1093/ntr/ntx109.[Epub ahead of print]).

[10] Hecht SS. Approaches to cancer prevention based on an understanding of N-nitrosamine carcinogenesis. Proc Soc Exp Biol Med. 1997 Nov;216(2):181-91.

[11] Hecht SS. Lung carcinogenesis by tobacco smoke. Int J Cancer. 2012 Dec 15;131(12):2724-32.

[12] AshleyDL, O'Connor RJ, Bernert JT, Watson CH, Polzin GM, Jain RB et al. Effect of differing levels of tobacco-specific nitrosamine-s in cigarette smoke on the levels of biomarkers in smokers. Cancer Epidemiol Biomarkers Prev.2010 Jun;19(6):1389-98.)

[13] Stephens WE, Calder A, Newton J. Source and health implications of high toxic metal concentrations in illicit tobacco products. Environ Sci Technol. 2005; 39:479-88.

[14] Tobacco, Euromonitor International (http://www.euromonitor.com/tobacco); International Tobacco Control Policy Evaluation Project (http://www.itcproject.org/resources/reports/2/); Latest, the Tobacco Atlas (http://www.tobaccoatlas.org/); Tobacco Control Country

profiles, WHO Tobacco Free Initiative. http://www.who.int/tobacco/surveillance/policy/country_profile/en/ All accessed 15 January 2018.

[15] Canada. Tobacco Reporting Regulations. (https://www.canada.ca/en/health-canada/services/health-concerns/tobacco/legislation/federal-regulations/tobacco-reporting-regulations.html, accessed 15 January 2018).

[16] Brazil. Resolution–RDC No.90,of December 27,2007.(http://www.tobaccocontrollaws.org/files/live/Brazil/Brazil%20-%20RDC%20No.%2090.pdf, accessed 15 January 2018).

[17] Harmful and Potentially Harmful Constituents in Tobacco Products and Tobacco Smoke; Established List. A Notice by the Food and Drug Administration.2012.(https://www.federalregister.gov/documents/2012/04/03/2012-7727/harmful-and-potentially-harmful-constituents-in-tobacco-products-and-tobacco-smoke-established-list).

[18] Djordjevic MV, Fan J, Bush LP, Brunnemann KD, Hoff-mann D. Effects of storage conditions on levels of TSNAs and N-nitrosamino acids in U.S. moist snuff. J. Agric. Food Chem. 1993; 41:1790-1794.

[19] Andersen RA, Fleming PD, Hamilton-Kemp TR, Hildeb-rand DF. pH changes in smokeless tobacco undergoing nitrosation during prolonged storage: Effects of moisture, temperature, and duration. J. Agric. Food Chem. 1993;41:968-972.

[20] WHO Tobacco Laboratory Network (TobLabNet). WHO TFI, Geneva. http://www.who.int/tobacco/global_interaction/toblabnet/en.

[21] WHO Tobacco Laboratory Network (TobLabNet) history. WHO TFI, Geneva. http://www.who.int/tobacco/global_interaction/toblabnet/history/en/index1.html

[22] Best Practices in Tobacco Control - Regulation of Tobacco Products Canada Report, footnotes 3 and 5 WHO Study Group on Tobacco Product Regulation (TobReg). 2005. http://www.who.int/tobacco/global_interaction/tobreg/Canada%20Best%20Practice%20Final_For%20Printing. pdf

[23] Canada. Tobacco Reporting Regulations (SOR/2000-273), Justice Laws website, current to 27-12-11. (http://laws-lo-is.justice.gc.ca/eng/regulations/SOR-2000-273/page-2. html#h-2, accessed 15 January 2018).

[24] The scientific basis of tobacco product regulation. Second report of a WHO study group. Geneva, 2008. (http://apps. who.int/iris/bitstream/10665/43997/1/TRS951_eng. pdf?ua=1&ua=1).

[25] ISO/IEC 17025:2005. General requirements for the competence of testing and calibration laboratories. International Organization for Standardization, Geneva. (https://www.iso.org/standard/39883.html).

[26] Westgard JO. A Total Quality-Control Plan with Right-Sized Statistical Quality-Control. Clin Lab Med. 2017 Mar;37(1):137-150.

[27] Collins G. Tobacco giant recalls 8 billion faulty cigarettes. New York Times. 27 May 1995. http://www.ny-times.com/1995/05/27/us/tobacco-giant-recalls-8-billion-faulty-cigarettes.html

[28] Publications, WHO Tobacco Free Initiative website. http://www.

who.int/tob-acco/publications/en/
- [29] Designation, WHO Tobacco Free Initiative website. http://www.who.int/tobacco/global_interaction/toblabnet/designation/en/
- [30] Participation and conflict of interest requirements, WHO Tobacco Free Initiative website. http://www.who.int/tobacco/global_interaction/toblabnet/conflict_of_interest/en/.

附录　实验室内和实验室间验证

A1.1　实验室内方法验证

所有的实验室方法都必须经过仔细的测试并确定它们满足如何使用数据的要求。由于不同的用途可能有不同的要求，因此不同的设备需要、数据要求应该在实验室决策过程之前进行评估。

准确性是测量值与量的真值的接近程度。准确性主要受到系统误差或偏差的影响。分析测量可能具有分析偏差，因此所确定的结果可能高于或低于实际值。准确性通常通过测试已知值的材料水平来评估，或者测试不同的测试方案之间的紧密度来评估，这些方案是独立的，不应该有相同的偏差。在这两种情况下，确定的值越接近已知的水平或一致的水平，考虑的特定测量就越精确。缺乏准确性不能通过更多的测量来克服。在一个常见的类比中，如果一连串的箭射向目标，准确度是各箭头与目标中心的距离的平均值。

精度是指如果在同一个样本上重复测量，通常使用相同的方法，测量结果之间的差距。精度主要受随机误差的影响，导致结果不一致。通过对相同的材料进行多次测量，然后在统计上确定结果的可变性，进行精确的评估。由于精度主要由随机误差决定，所以在某种程度上，可以通过进行更多的测量来解决精度不足对测量精度的影响。采用的测量值越多，这些测量值的平均值就越接近使用该方法确定的值。

但计算精度可受其影响。该方法包含在精度测定过程中。例如，在完成样本制备步骤后，用重复仪器分析确定的结果通常比通过样本制备步骤和分析测量过程确定的样本上的数据更具有可再现性；分析方法的两个部分都可能引入随机误差。在上面同样的类比中，如果一系列的箭射向一个目标，那么精度将是各箭彼此之间的接近程度，即使组合在一起的平均数不接近目标的中心。

灵敏度是在低水平下做出精确和准确测量的决定能力。换句话说，它是分析系统在分析物存在时检测分析物的能力。灵敏度受到整个分析过程的影响，包括分析物提取、净化、浓度和分析。灵敏度可以表示为检测限，通常定义为重复测量空白样本的三倍标准偏差。或者，也可以定量限来描述，定量限是重复测量的标准偏差的10倍。灵敏度可以通过一些或所有分析步骤的改进而提高；仪器的升级可以提供实质性的改进。

灵敏度的一个伴随概念是选择性，或者说特异性。选择性是指当一种物质确实不存在时，正确地识别它不存在的能力。选择性主要受到样品中存在的污染物的影响，这些污染物的性质足够接近目标分析物而不能与目标分析物区分开来。选择性通常可以通过更好的样品制备方法和更先进的仪器来改进。

稳健性是分析系统能够承受偏离所定义的分析方法的能力。偏差可以包括各种各样的现象，从称量材料的错误到仪器操作从一个维护操作到下一个维护操作的变化。为了正确评估，应该评估最可能的偏差，以了解这些偏差如何影响最终结果。适当的稳健性评估将确定那些对测量影响最大的方面，并且应该被最密切地监视。

A1.2　实验室间方法验证

如果一个实验室的数据要与另一个实验室的数据进行比较，那么实验室间的验证是很重要的。此外，如果一个实验室想要确定他们的结果与其他实验室确定的结果一致，那么实验室间的验证是必不可少的。实验室间验证的评估有几个可用的方案。在 ISO 5725-1 和 ISO 5725-2 中有一种广泛接受的数据评估方法[1,2]。实验室间验证过程由一个单一来源组成，为一系列实验室提供等效的样品。这些样本在每个实验室的操作条件下使用不同的方法进行分析，并反馈结果。然后对所有结果进行评估，以确定结果的可重复性。根据定义，在最短的可行时间内对匹配的卷烟样品产生两个结果的差异不超过重复性，同一操作员使用同一仪器，其平均值在正确应用此方法条件下不超过二十分之一。两个实验室报告的匹配卷烟样品的单个结果的差异会超过再现性，但也在正确用此方法的条件下平均频率不超过二十分之一。与烟草科学研究合作中心 (CORESTA) 进行了几次实验室间的验证，以支持烟草工业的努力[3,4]。世界卫生组织的 TobLabNet 也进行了一系列实验室间验证，如第 3.2 节所述。

参考文献

[1]　ISO 5725-1:1994.Accuracy (trueness and precision) of measurement methods and results -- Part 1: General principles and definitions. International Organization for Standardization, Geneva. (https://www.iso.org/stan-dard/11833.html).

[2] ISO 5725-2:1994. Accuracy (trueness and precision) of measurement methods and results -- Part 2: Basic method for the determination of repeatability and reproducibility of a standard measurement method. International Orga-nization for Standardization, Geneva. (https://www.iso. org/standard/11834.html).

[3] Intorp M, Purkis, S, Wagstaff, W, 2010a. Determination of aromatic amines in cigarette mainstream smoke: the CORESTA 2007 Joint Experiment; Beiträge Tabakforsch. Int. 24(2) (2010) 78-92.

[4] Intorp M, Purkis, S, 2010b. Determination of selected volatiles in cigarette mainstream smoke. The CORESTA 2008 Joint Experiment; Beiträge Tabakforsch. Int. 24(4) (2014) 174-186.)

Preface

It is well established that tobacco use is a major public health problem. However, tobacco products are one of the few openly available consumer products that are virtually unregulated in terms of contents, design features and emissions. The majority of countries hesitate to implement regulations in this area, partly due to the technical complexity associated with tobacco product regulation. There has been a high demand from WHO Member States for resources consolidating information on tobacco testing and building laboratory capacity for countries, especially to facilitate the implementation of Articles 9 and 10 of the WHO FCTC[1]. This is to provide a useful, comprehensible and easy guide for regulators and policymakers on how to test tobacco products, what products to test, and how to use testing data in a meaningful way to support regulation.

The importance of laboratory testing is reflected in the WHO Framework Convention on Tobacco Control (WHO FCTC). Article 9 of the WHO FCTC defines obligations for Parties with respect to the testing of tobacco products, while Article 10 deals with the disclosure of information on the contents and emissions of tobacco products. The disclosure of

---

1 Participants of a WHO workshop on the How-to's of Establishing a Testing Laboratory in (April 2016, New Delhi, India) requested WHO to prepare a handbook on building laboratory capacity. Additionally, the WHO Tobacco Laboratory Network's sixth meeting (Maastricht, Netherlands, 9-11 May 2016) recommended the development of a primer informing governments and the public of WHO TobLabNet's activities in order to expand membership and build testing capacity globally.

TOBACCO PRODUCT REGULATION
Building laboratory testing capacity

product information takes two forms: 1) the disclosure of information by manufacturers to regulators, and 2) the disclosure of information from regulators to the public. Tobacco product testing is used to generate data necessary to support both forms of disclosure.

In 2006, the first Conference of the Parties (COP) to the WHO FCTC established a working group to elaborate guidelines and recommendations for the implementation of Articles 9 and 10 of the Treaty (Decision FCTC/COP1(15)). COP 2 extended the mandate of the working group and encouraged WHO's Tobacco Free Initiative (WHO TFI) to continue its work on tobacco product regulation (Decision FCTC/COP2(14)). In 2010, the partial guidelines submitted at COP4 were adopted. The partial guidelines currently contain recommendations for regulation to reduce the attractiveness of tobacco products. Recommendations to reduce the addictiveness and toxicity of tobacco products will be developed later. The working group was requested by the COP to continue its work to elaborate the guidelines in a step-by-step process, with updates on addictiveness and toxicity requested to be submitted to future sessions of the COP for consideration.

It is important to note that, contrary to claims by the tobacco industry, these guidelines are final and in effect. The regulatory measures advocated by the partial guidelines are to be treated as minimum requirements and do not prevent Parties from adopting more comprehensive measures.

WHO has continually supported Member States in developing laboratory capacity. In 2004, WHO TFI published a recommendation from the WHO Study Group on Tobacco Product Regulation (TobReg) on 'guiding principles to increase laboratory capacity to facilitate the implementation of Articles 9 and 10 of the WHO FCTC and to guide the initiation of tobacco product testing'. *(1)* The guiding principles provided advice to countries intending to develop such capacity and help in realising this objective. Over the intervening years, new knowledge has developed and progress has been made to support these efforts; these include establishing the WHO Tobacco Laboratory Network (TobLabNet) in 2005 and the Global Tobacco Regulators Forum (GTRF) in 2016. Therefore, it is appropriate to update the previous document and provide a practical guide for countries interested in developing or accessing tobacco product testing capacity to support their regulatory authority.

This document provides options for building laboratory capacity, which include developing a testing laboratory, using an existing internal laboratory, contracting an external laboratory, and making use of the support mechanisms available, including but not restricted to WHO TobLabNet. Finally, it provides practical, step-by-step approaches to implementing tobacco testing and is relevant even to countries with inadequate resources to establish a testing facility.

Acknowledgements

Main contributors to development: David L. Ashley, Ph.D (RADM (retired), US Public Health Service, Limited-Term Professor, Division of Environmental Health, School of Public Health, Georgia State University), and staff from WHO Department of Prevention of Noncommunicable Diseases.

Thanks to Nuan Ping Cheah, Ph.D (Director, Pharmaceutical, Cosmetics and Cigarette Testing Laboratory, Health Sciences Authority, Singapore/Chair, WHO Tobacco Laboratory Network), Dr. Ghazi Zaatari, (Professor & Chair, Department of Pathology & Laboratory Medicine, American University of Beirut, Lebanon/Chair, WHO Study Group on Tobacco Product Regulation), and the WHO FCTC Secretariat, for reviewing the text and providing useful comments.

Funding for this publication was made possible, in part, by the Food and Drug Administration through grant RFA-FD-13-032.

Glossary

Accreditation — the documentation by an independent body that a laboratory has the systems in place that should enable them to produce reliable results that have been adequately tracked and verified.

Accuracy — the nearness of a measurement of a quantity to the quantity's true value

CDC — U.S. Centers for Disease Control and Prevention

DAD — diode array detector

FID — flame ionization detection

Firewall — a system to ensure that data and information are protected so that public health and commercial interests are separate and not accessible to each other

GC — gas chromatography

HPLC — high-performance liquid chromatography

Labstat — a private commercial tobacco analysis laboratory, Labstat Incorporated, in Kitchener, Ontario, Canada

LC — liquid chromatography

MS — mass spectrometry

MS/MS — tandem mass spectrometry

NCEH — National Center for Environmental Health at the U.S. Centers for Disease Control and Prevention

OSH — Office on Smoking and Health at the U.S. Centers for Disease Control and Prevention

PAHs — polynuclear aromatic hydrocarbons are multi-ringed aromatic compounds varying from two rings (naphthalene) to much larger ringed structures (e.g. Indeno[1,2,3-c,d] pyrene)

Precision — a determination of how close measurement results are to each other if a measurement is made repeatedly on the same sample, typically using the same method

Quality control — a process which evaluates whether systems are operating within standard parameters on an ongoing basis

Ruggedness — ability of an analytical system to withstand deviations from the defined analytical method.

Selectivity — the ability to correctly identify that a substance is not present when it is indeed not present.

Sensitivity — the ability of a measurement to make accurate and precise determinations at low levels.

TCD — thermal conductivity detector

TFI — Tobacco Free Initiative of the World Health Organization

TobLabNet — WHO Tobacco Laboratory Network

TobReg — WHO Study Group on Tobacco Product Regulation

TSNAs — tobacco-specific nitrosamines [N-nitrosonornicotine (NNN), 4-(N-nitro- somethylamino)-1-(3-pyridyl)-1-butanone (NNK), N-nitrosoanatabine (NAT), and N-nitrosoanabasine (NAB)]

UV — ultraviolet

WHO — World Health Organization

WHO FCTC — WHO Framework Convention on Tobacco Control

Chapter 1   Testing in the Context of a Country's Regulatory Authority

Tobacco product testing is a valuable tool to support tobacco control and regulatory efforts, which can have a clear impact on population health. Tobacco product testing per se does not lower the levels of toxic and carcinogenic constituents in tobacco, or reduce the use of tobacco products which cause the exposure of both users and non-users to the harmful chemicals in tobacco product emissions. However, it can be an effective tool if the objectives and justifications are set out upfront on how data on design, contents or emissions will be used for regulatory purposes. How that tool is used is a critical factor in determining whether tobacco product testing is effective.

It should be understood from the start that obtaining data on design, contents or emissions alone is not an adequate reason to require tobacco product testing. It is critical that countries establish sound reasoning on how this information will be used because testing can be expensive, even if the manufacturers are required to fully fund the work, and a solid justification is important in order to defend against legal attacks or address legitimate questions posed by government officials. For the purposes of this document, the term "manufacturers" will be used to refer to tobacco product manufacturers, importers, or other companies that fall under a country's tobacco product regulatory authority and are responsible for marketed tobacco products.

The first step in developing a tobacco product testing programme is

TOBACCO PRODUCT REGULATION
Building laboratory testing capacity

identifying the basis and justification for testing and reporting. In addition to providing an important justification, these considerations will point the programme in a direction that will be most beneficial and ensure that the effort provides data that can be used to effectively support tobacco control and regulation in that country. While there are some good examples of how other countries have approached the need for product testing, each country's experience is different. Identifying how information will be used at the national level is the foundation upon which all further action is built and the unique needs of the country must be the first consideration.

There are two ways to use laboratories in a tobacco regulatory scheme, depending on the way in which government regulatory agencies intend to require tobacco product testing. The first approach is for the government agency to oversee or carry out routine analysis of all products. Even in this case, requirements can be put in place so that the cost of these analyses are borne completely by the manufacturers (see Section 1.7 below), but this effort would require a significant logistical effort that many countries may not want to undertake. Alternately, countries may choose to limit their direct involvement in testing to assessing the accuracy of data reported by manufacturers, using a random scheme for choosing a subset of brand/subbrands and analytes, and reanalysis by the government itself, or having other mechanisms in place to ensure this accuracy. As explained above, manufacturers can be required to fund this testing. This document is applicable to both approaches.

1.1 Initial considerations

The first step in identifying what testing should be required is to carefully

Chapter 1 Testing in the Context of a Country's Regulatory Authority

evaluate the government's powers to require testing. Every country's authorizing legislation is different and these differences must be considered when determining how best to use the testing data that will be obtained. For many countries, national tobacco control legislation can be guided by the treaty obligations under the WHO Framework Convention on Tobacco Control (WHO FCTC). Articles 9 and 10 of the WHO FCTC and their partial guidelines, set out the recommendations for Parties to the Treaty to adopt, while framing national legislation on tobacco product regulation.

1.1.1 Reason for the regulatory authority

The first question to ask is the primary purpose for creating the tobacco product regulatory authority. There are several reasons as follows:
- One common concern relates to the harm to children's health resulting from taking up tobacco product use. While many health consequences from tobacco use take a long time to manifest, the testing of tobacco product ingredients and emissions that have a particular appeal to children may constitute the highest testing priority.
- Other ingredients or design features could also be of high priority for analyses and would determine early decisions about such testing capability.
- The legislation may have been based on the rights of non-users not to be exposed to harmful chemical constituents. In this case, measurement of harmful emissions in second-hand smoke may be the most critical data from tobacco product testing. Measurement of chemical agents in second-hand smoke is more

challenging than measurements in mainstream smoke, so, if this is a critical issue, laboratory decisions should consider the ability of laboratories to make these measurements.
- Another possible issue may be false advertising and other marketing statements by the tobacco industry. In this case, it is critical that testing capabilities be able to verify such claims.
- If there are claims that one product poses a lower risk, the ability of product testing to address these claims should be a central consideration.

There are other issues which may have been the primary consideration of legislation and these should also be considered when making testing decisions.

1.1.2 Scientific basis of the legislation

Another equally important aspect to consider is the scientific basis for the legislation. As decisions are made concerning what laboratory capability is necessary, it is important to consider whether the original decision and rationale were strongly supported by scientific data, whether it was driven by public health policy concerns or whether it was based on treaty obligations. If legislation was particularly based on scientific data, laboratory data that supports the legislation may be part of a critical defence of the decision taken by the regulatory authority. In this case, the data derived from laboratory testing must be unimpeachable. Tobacco regulatory authorities must ensure that the public release of data will strengthen legal arguments. In most cases in which there is a strong scientific rationale for regulating tobacco products, the choice

Chapter 1 Testing in the Context of a Country's Regulatory Authority

of a well-established, experienced and accredited laboratory will prove indispensable in how the data is eventually used.

1.1.3 Public health concerns

From a public health viewpoint, regulatory agencies should consider identifying the major issues of concern. This will differ from country to country, and it is important that regulatory agencies determine what concerns are most important and can be addressed using laboratory data. For example, does the use of manufactured cigarettes cause the biggest tobacco-related public-health problem? This may be the case in many countries, but others may have a bigger public health concern about other tobacco products, such as the various forms of smokeless tobacco, waterpipes, *bidis*, *kreteks* or flavoured products. If these other products are the biggest public health problem, testing of manufactured cigarettes may be the most straightforward, but not the most effective choice for tobacco product testing, to aid effective regulatory actions. Another public health concern may be the introduction of new products. Whilst the overall impact of new products might at first be unclear, tobacco regulators will likely be asked about them. The ability to provide scientifically-based answers will demonstrate the usefulness of the testing capability. Focusing testing resources on the real source of public health concern will demonstrate a better rationale for requiring the testing and more readily achieve real improvements in public health.

Tobacco regulatory authorities should also identify the real objectives of the regulatory programme and how tobacco product testing can help. The specific objectives may be to ban certain products from the market, reduce prevalence of use, or reduce disease and death resulting from tobacco product use. Too often, decisions are based on what appears to be most readily accomplished and not whether this fits into an overall goal for tobacco regulation. When tobacco product testing is aimed at supporting the main strategic objective, it demonstrates its value and leads to better results.

1.1.4 Resonance with the public and decision-makers

It is also important to consider the data that would most resonate with the public and the authorities, and how such data can be made comprehensible. This is important because, as discussed above, tobacco product testing is only effective when coordinated with and used by the tobacco control or regulatory authority. For example, if addiction is a major public health issue, measurement of nicotine, the primary addictive component of tobacco products, may be the primary aim for laboratory capability. If the toxicity of the product and its impact on causing disease is the main issue, this would point to the measurement of toxic and carcinogenic substances, such as tobacco-specific nitrosamines (TSNAs) and polynuclear aromatic hydrocarbons (PAHs) in emissions. If tobacco control authorities need data showing that non-users are being exposed to second-hand smoke, this may be the most effective way to accomplish public health goals. If the data addresses a public health issue that is clear to the public and decision-makers, or can be made clear, the data is more likely to be used to substantially improve public health.

1.1.5 Legal requirements

Finally, a country's legal requirements and its possible impact on

Chapter 1 Testing in the Context of a Country's Regulatory Authority

enforcement must be considered. Lack of certification or adherence to standards may impede admissibility, or reduce the weight to be attached to a laboratory report in court proceedings. Issues like chain-of-custody or participation in inter-laboratory validation (see Appendix 1) could be important considerations in the choice of laboratories to carry out tobacco product testing. It is most likely that tobacco manufacturers will use highly qualified laboratories and any compliance or enforcement proceedings taken against a manufacturer would need to match the qualifications of the scientific source to be credible. Government bodies should consider relevant rules of evidence in selecting laboratories to assist in enforcement and other proceedings. They may also work in collaboration with the ministry of justice, or some other judicial or legal administration, while developing the testing requirements to ensure that laboratory data are adequate as evidence for legal proceedings.

1.2 How data can be used

Laboratory data provide an opportunity to avoid statements of opinion or anecdote by offering statements of fact that have a definitive basis in scientific determination. Without strong scientific data, the grounds for regulatory action will be more easily questioned and rebutted, and any action taken is less likely to accomplish its objective. While laboratory data do not guarantee success, scientific data strengthen every rationale and increase the likelihood that goals will be achieved.

Enforcement authority, which can be used to reduce disease and death from tobacco product use, differs from country to country. But the powers granted by a national government carry both risks and opportunities.

TOBACCO PRODUCT REGULATION
Building laboratory testing capacity

Unfortunately, most tobacco regulatory agencies do not have adequate resources to make use of all of the data that might be made available. Stretching an organization too thin makes it less likely that goals will be achieved. So, it is critical that regulators assess what can be done with the powers and resources available and how to use them most effectively. Identifying how the data will be used should determine what data are collected and what requirements are placed on that data.

Some possible ways that countries might use testing data include the following:
- setting product standards
- limiting advertising claims
- educating the public
- informing future legislation
- marketing authorization
- developing scientific information to support research, and
- setting manufacturing standards.

1.2.1 Product standards

When a country has the regulatory authority, setting product standards can be a valuable regulatory tool in reducing disease and death resulting from tobacco product use. But because they can be so effective, these standards are likely to be challenged in the courts. To prepare for these challenges, supporting evidence must be derived from well-documented and peer-reviewed scientific evidence. Product standards used by national regulatory bodies include those aimed at both the addictiveness and attractiveness of tobacco products.

Chapter 1   Testing in the Context of a Country's Regulatory Authority

- In 2012, Brazil enacted a product standard banning the use of additives in cigarettes and other tobacco products sold there. *(2)* This was based on the impact of additives to encourage the use of products by young people and facilitating initiation of their use.
- In 2009, Canada enacted a ban on flavours, except menthol, in cigarettes and cigarillos. *(3)* The action was intended "to protect the health of Canadians",
- "to protect young persons and others", and "to enhance public awareness of the hazards of tobacco use". This ban was recently expanded to include menthol. *(4)*
- In 2009, France also adopted a law restricting the use of flavouring ingredients in cigarettes in an effort to reduce youth initiation of smoking. *(5)*

All these standards were based on laboratory data that demonstrated the presence of ingredients of concern in tobacco products. The choice of product standards should be driven by critical country-specific issues as discussed above.

1.2.2   Marketing/Advertising restrictions

Marketing/advertising restrictions have been used by several countries to make products less appealing. This can take the form of limiting direct marketing, such as statements of lower risk for certain products or indirect marketing appeal, such as the use of colours and imagery in packaging. For example, in 2001, Brazil was the first country to ban misleading terms, such as "light" and "low-tar" on tobacco product labelling. *(5)* While laboratory data alone is unlikely to be sufficient to support action on marketing/advertising restrictions, it can be used as a factor in evaluating relative risk statements and to support action banning or restricting these statements.

1.2.3 Public education

Public education may be one of the most effective applications for using testing data by countries just beginning to establish a tobacco regulatory programme. The public, in general, is not scientifically knowledgeable. Users and non-users alike do not understand how the design, contents and emissions of tobacco products affect their health. Many do not understand that increases in exposure to harmful chemicals results from the process of growing, manufacturing and use of tobacco products. *(6, 7)* Improving that understanding with information from product testing can be a valuable way to inform users and discourage tobacco use. But it is important that information provided to the public is scientifically sound. Trust in the reliability of the government agency to provide accurate information is critical to countering false messages provided by tobacco product manufacturers.

Public education can be a valuable tool to help users make informed choices and for non-users to be aware of the dangers associated with exposure to emissions from tobacco products. This education can take many different forms. Experience in Canada has shown that the public has a weak understanding of numeracy and can be misled by placing machine-derived numbers on products. The Regulatory Impact Analysis Statement that was published with the amendment to the Tobacco Products Information Regulations (TPIR) states:

> Research has shown that the current format of the toxic emissions statements, which displays a range of values for six toxic substances,

Chapter 1 Testing in the Context of a Country's Regulatory Authority

is generally not noticed by tobacco users and many people find them confusing. The proposed Regulations would replace the numerical values currently displayed with four text-based statements that provide clear, concise and easy-to-understand information about the toxic substances found in tobacco smoke. *(8)*

On the other hand, the public generally wants to avoid exposure to "chemicals" especially when these can be linked to adverse health outcomes. *(9)* So while care must be taken on how to make available information on toxic and addictive substances in tobacco products, this should be provided to the public in a meaningful manner.

1.2.4 Informing future legislation

It is likely that most countries' legislation to enact treaty obligations under the WHO FCTC was not comprehensive. Data derived from product testing provide information for future legislation. This future legislation could take many different forms and should be carefully considered based on the overall purposes of the tobacco regulatory programme, as discussed above. If a product standard authority was not included in the original legislation, testing of tobacco products may be a valuable source of data demonstrating its value. Another example of additional legislation may be complete smoking bans in indoor public places, workplaces and public transport, as enacted in many countries in recent years. Since indoor smoking bans are largely driven by concern for non-smokers' exposure to second-hand smoke, the focus of this effort would be on the chemicals to which non-smokers are exposed. Product testing can identify and quantify the toxic and addictive chemicals that are emitted from using tobacco products to which these non-users are exposed.

1.2.5   Notification or marketing authorization

Some countries receive notification when a new product is introduced to the market, with some regulatory agencies specifying notification requirements in their national legislation or tobacco control laws. As part of this process, tobacco manufacturers may be required to provide detailed product information, including the ingredients used in the tobacco products marketed in the burnt and unburnt form, quantities thereof, their toxicity, as well as possible adverse effects. The use of testing to evaluate these products and ingredients can be a valuable use of the testing and reporting mandate under the WHO FCTC.

Most countries do not have authorization to determine, before marketing, whether a product can be authorized for sale. For those countries that do, this can be a very powerful regulatory authority, but it requires substantial internal resources. When that authority has been granted, product testing is an important factor in evaluating marketing authorization. Because companies want to be allowed to market their products, they are willing to provide a wealth of testing data as required. To properly use this data, regulatory authorities must have the dedicated scientific resources and expertise needed to evaluate data provided by the manufacturers.

1.2.6   Developing scientific information

Data generated from testing of tobacco products can be used to develop scientific information which may be valuable for others, such as researchers, to better understand the impact of the country's tobacco products on disease and death. Because of limited resources, issues

Chapter 1   Testing in the Context of a Country's Regulatory Authority

related to trade secrets and the impact of regular design changes on tobacco product emissions, researchers are rarely able to fully evaluate the tobacco products that their test subjects are using. If this information can be obtained from the manufacturers and made available to researchers and other interested parties, it would be a valuable tool to improve the interpretation of the results of human research into a specific country's tobacco product use. These data can then be used to support many of the purposes listed above, including shaping future regulations.

1.2.7   Manufacturing standards

Manufacturing standards are another regulatory tool to address the harm caused by the use of tobacco products. Testing data can identify the variability of levels of chemical ingredients in production starting materials. For instance, tobacco-specific nitrosamines (TSNAs) are some of the most potent carcinogenic agents in tobacco products. *(10, 11)* But the levels of nitrosamines delivered to the user are highly dependent on the levels of carcinogens in the original tobacco used in manufacture and these levels vary widely depending on the tobacco. *(12)* By setting limits on the levels of chemical constituents in ingredients used in manufacturing or contaminants, such as heavy metals *(13)*, exposure of product users can be reduced. But these levels must first be determined by valid testing of tobacco products.

1.3   Identifying tobacco products to test

Each country will have products that are best targeted for regulation. If resources to evaluate testing results are limited, which is generally the case,

TOBACCO PRODUCT REGULATION
Building laboratory testing capacity

it may be best to identify highly significant products and consider these as a testing priority. The following discussion is not intended to eliminate any particular product from the testing requirement, but to suggest factors that regulators may consider when determining which products should be the highest priority.

There are three factors to be considered in evaluating which product to address as the highest priority:

1. which types of tobacco products are most prevalent in the market
2. which types of tobacco products are the most harmful to users
3. which tobacco products are most feasible to regulate?

1.3.1 Types of tobacco products most prevalent in the market

In order to have a substantial impact on reducing disease and death from tobacco use, it is important to address the type of product which has, or is likely to have, a large market share . If manufactured cigarettes are only used by a small fraction of the population, even major reductions in use will only have a minor public health impact. There are several sources of data that can be used to evaluate the prevalence of use of different types of tobacco products in a country. *(14)* Global surveys and similar efforts have identified the number of people using different types of tobacco products. Many of these surveys also break down use by gender and age. The prevalence of product use varies dramatically between countries. In Indonesia, *kreteks* (flavoured cigarettes) are prevalent, but these products have only a minor market share in most other countries. Smokeless tobacco use in India is very common, but the smokeless products being used are very diverse. Another example is the use of menthol cigarettes,

Chapter 1   Testing in the Context of a Country's Regulatory Authority

which is widespread in some countries, such as the Philippines. So focusing regulatory product testing efforts on products other than manufactured cigarettes may be the best use of resources for some countries.

1.3.2   Types of tobacco products most harmful to users

Toxicity and the harm caused by tobacco products differ both within and between product classes. It is generally understood that, because of the high concentrations of very toxic and carcinogenic chemicals delivered to the lungs, combusted traditional products (i.e., cigarettes, cigars, *bidis*, *kreteks*, waterpipes, etc.) pose the most harm to users. The diversity of smokeless tobacco products poses further regulatory and testing challenges. But the toxicity of various products could vary depending on specific manufacturing practices and user behaviour. A product that is less toxic but used often may be a bigger concern than one that is more toxic but only rarely used. When choosing the tobacco products on which efforts should be focused, regulators should consider which products in their market present the most significant threat to health.

1.3.3   Tobacco products most feasible to regulate

Regulatory opportunity is the third factor to consider when identifying the tobacco products on which to focus initial product testing requirements. Depending on the specific authority given by the enacting legislation and the nature of the political climate, some actions may be easier to accomplish than others. As indicated above, new products being introduced into the marketplace that do not have a significant market share may be a more viable initial target than well-established products with strong stakeholder support. Generally, products that are manufactured in a limited number of facilities and not substantially altered by the user are more readily regulated than products made by a cottage industry. When products are made by hundreds of thousands of small manufacturers, enforcement of required testing could be so challenging that this should not initially be the highest priority when establishing new testing and reporting requirements. Regulators should consider the feasibility of successfully regulating an industry that is very widespread as part of their prioritization process. An example of this concern would be *bidi* manufacturing in India. *Bidis*, hand-rolled tobacco products, are widely used and present a significant health concern to users. But there is a large manufacturing sector for these products in private homes or very small shops. Enforcing testing requirements for these cottage-industry products in India would be very challenging and might not be the first priority for testing. After successes with other products, *bidis* may be later identified as a target. Also, the usefulness of data from manufacturers of tobacco products that are altered by consumers (e.g., adding lime to increase free nicotine levels) should be considered as part of the process for determining the products for which testing and reporting data should be the highest priority.

1.4   Analytes to test

Several countries have developed lists of analytes – chemical substances measured using chemical analysis – to be tested. Canada was one of the first countries to identify lists of analytes to measure in mainstream smoke, sidestream smoke and whole tobacco. *(15)* In 2007, Brazil also established

Chapter 1 Testing in the Context of a Country's Regulatory Authority

a list of design properties, contents and emissions *(16)* to be tested. In 2012, the US Food and Drug Administration (FDA) published a list of 93 harmful/potentially harmful constituents in tobacco products and tobacco smoke. *(17)*

These lists can serve as a starting point for countries which intend to require testing and reporting of design properties, contents and emissions from tobacco products. But the decision concerning which analytes should be tested involves several factors and should be carefully considered.

The first factor is which analytes best meet the purpose of how data is intended to be used, as described in Section 1.2 above. Analytes should be chosen using a ratio-nale to link the results of testing data with their use. For example, if the regulatory agency intends to communicate to the public or decision-makers the concern for cancer-causing chemicals in cigarette smoke, clear choices for testing would be TSNAs and PAHs (or benzo[a]pyrene as a surrogate for PAHs), since they are known carcinogens and have been linked to cancer in tobacco product users. Alternatively, testing of heavy metals for the purpose of setting product standards may not be a good choice, since heavy metals in tobacco are largely driven by levels in the soil in which the tobacco is grown and not by the manufacturing process. It is important to know if this is an issue in the products being marketed and if a standard can be used to reduce these levels. How testing data is to be used is a critical factor in determining which analytes to measure.

An equally important factor is which analytes are of most concern in the products identified in Section 1.3. The use of certain products and the health effects from their use should be considered when identifying the analytes to be measured in tobacco product contents and emissions. For example, measurement of carbon monoxide is an analyte of concern in combusted tobacco product emissions. But because smokeless tobacco is not burned when used, requiring measurements of carbon monoxide in traditional smokeless tobacco is both unnecessary and inappropriate. It is important that the analytes chosen be relevant to the products to be tested. Testing of constituents that are generated only by the burning of tobacco is not appropriate for testing of traditional smokeless tobacco products.

Thirdly, those analytes to be tested should have reliable methods for their determination. For the purposes of this document, reliability is considered to include the ability to produce accurate and reproducible results at appropriate detection limits with suitable sensitivity and selectivity. When first establishing a list of analytes to be measured in tobacco product contents and emissions, analytes that have already established, widely accepted, and sensitive analytical methods would be the highest priority so that results can be obtained as quickly as possible. Analytes for which no established methods exist can be added at a later date when the testing programme is more mature and the value of testing has been established.

1.5 How the tests are to be conducted

How analytes are tested is another important consideration when developing the requirements for laboratory testing. Different countries have approached this in different ways and there are limitations based on national laws and acceptable requirements. One issue is whether to allow the use of different analytical methods to measure the analytes in tobacco product contents or emissions, or require the use of specific methods. The requirement to use specific methods, as is done in Canada, has certain

Chapter 1 Testing in the Context of a Country's Regulatory Authority

distinct advantages. One of the issues that will arise in tobacco product testing is data comparability at one time and over time. Data generated using different methods are not always comparable, even though they should be, due to differences in accuracy, sensitivity and selectivity. These issues are largely overcome when the same method is used because these differences are largely removed. But they may not be fully addressed because of inter-laboratory differences in carrying out these measurements (see Section 4.3). Inter-laboratory differences can be addressed to a large extent through participation in inter-laboratory comparison studies. But if possible, a better way to guarantee comparability is to require the use of the same analytical method by the same laboratory. This will maximize the comparability of data. WHO TobLabNet has developed and globally validated methods for testing of some priority tobacco products contents and emissions.

The disadvantages to this approach result from its rigidity. If the same method is required by legislation or regulation, it may be hard to adopt new more effective methods as science advances. The method that is specified will, over time, not keep up with the development of new and more sensitive, more reproducible methods which could benefit the interpretation of analytical data for public health purposes, at least until the legislation or regulation is updated. Also, the requirement to use a single laboratory, if allowed, would eliminate competition which can reduce costs and encourage the development of additional testing capability by initiating development of other labs. If allowed, the choice of methods and/or laboratories, the reliability of the method, and the reliability of the laboratory doing the testing must be fully evaluated before the data are accepted (see Sections 4.2, 4.3 and 5.3). This can require an extensive effort and needs expert advice. Requiring measurement of known standards can help. Regulatory agencies should consider these trade-offs when deciding which approach to take.

1.6 Communicating data to regulators

Requirements concerning the data to be communicated to regulators are as important as the choice of analytes and how they are to be tested. Regulators need to have information about how the measurements were made so that the quality and comparability of the data can be assessed and appropriate action taken. Analytical results alone, without the context of how they were determined, have limited use.

Regulators must decide the frequency of data to be reported on each brand/subbrand.

Acceptable frequency may range from twice a year to once every two years. More frequent measurements help evaluate variation between manufacturing runs but increase cost and the resources necessary to collect, compile and evaluate the data provided. The regulatory agency should weigh these factors before requirements are finalized.

There are some reporting requirements that are obvious. The undisputable identification of the product tested must be provided. This includes information on the brand/subbrand such that the specific product can be identified. As a minimum, subbrand information should include:
- the size of the article[2] (length and diameter for cigarettes)

---

[2] Article refers to the specific product used by the consumer. For example, for cigarettes, an article is the actual cigarette stick that is burned.

Chapter 1 Testing in the Context of a Country's Regulatory Authority

- the number of articles or size of the package (e.g., 20 cigarettes, 3 ounces for smokeless)
- ingredients added, including flavours (e.g., menthol, strawberry, mint)
- tobacco cut size for smokeless (e.g., long cut)
- ventilation level for cigarettes, and
- any other designator that a manufacturer or a consumer would use to distinguish between products of the same brand name.

It is also critical that the levels determined along with the units of measure (e.g., mg/cigarette, mg/gram of tobacco) be reported. Finally, all analytical determinations made, the number of replicates and the overall mean among those statistically accepted data should be reported. It is important that all results, even those that were rejected, be included along with the reason for rejection so that the data that were reported can be properly assessed. Regulators should also specify the number of significant digits (typically three) that should be reported. Differences that may be significant between samples (e.g. 3.12 versus 3.45 mg/g) could be lost if too few significant digits are required (e.g. 3 versus 3 mg/g).

Additional supporting information that establishes the quality of the analytical measurement is highly recommended. The report to the regulator should include the method(s) used to make the analytical measurements and the method validation parameters (see Appendix 1). To properly assess the reliability of the data and understand if it can be compared to other data being reported and to previous data reported to the regulator, or in the peer-reviewed literature, it is necessary to know the method(s) that were used and their accuracy, reproducibility, sensitivity, and selectivity. Only then can a proper comparison be made. Additional data to be reported to the regulator which help demonstrate the quality of the results reported include quality control results demonstrating that the analytical system was operating properly when the measurements were made, and levels of known standard materials. The measurements of samples, the levels of which have been independently established, can be used to evaluate whether the levels reported are in line with scientifically-accepted results, and by implication whether the results reported on unknown samples are valid.

It is also important to include information about how the testing samples were selected. In addition, the location(s) from which the sample(s) was/were taken (e.g., from the manufacturing line, storage room, retail location) and the shipping and storage conditions that the sample has experienced must be provided, as some analytes change under certain storage conditions. For example, under some storage conditions, the levels of TSNAs rise in some tobacco products. *(18)* Also, the pH of smokeless tobacco has been shown to change during storage, altering the levels of free nicotine. *(19)* In these cases, it is not appropriate to compare analytical results for samples stored for different times under different conditions. In order to reduce bias, the means of selecting samples should be specified. This could include a randomization scheme for samples that have been placed into the same storage room, a requirement for sampling from multiple manufacturing runs, blind selection of samples at retail, or other means. The randomization scheme should be designed so that the samples selected are representative of the products that are marketed. This is necessary to ensure that samples are not manufactured and sampled specifically for the analytical test but are representative of products sold to consumers.

Chapter 1   Testing in the Context of a Country's Regulatory Authority

1.7   Covering costs

Regulatory agencies and governments should not bear the cost of testing and reporting. The manufacturer should bear all costs as a condition of doing business and having access to markets. This is standard practice for most industries (food, cosmetics, pharmaceuticals) and should also be applicable to tobacco product manufacturers. Covering the costs of testing can be accomplished by a direct transaction between the manufacturer and the testing laboratory. Though they may choose to do so, it is not necessary for regulatory agencies to act as intermediaries, receiving samples from manufacturers and sending these off for testing. This imposes a substantial logistical burden on the regulatory agency that is best borne by the manufacturer. But to ensure data accuracy and integrity, other safeguards as described in this document should be instituted.

Regulators may incur some expenses for their part in the testing programme. These include evaluation of the data reported, oversight of the testing and reporting system, enforcement activities, and analyses to check the authenticity of the reported results. There are several mechanisms that countries might use to ensure that these costs are also borne by the manufacturers.

Some countries may choose to impose user fees on manufacturers to cover government regulatory costs and to allow the marketing of tobacco products. User fees can be based on the number of brands/subbrands for a particular manufacturer in the market, or the market share of a particular product. A set amount should be established for the functioning of the regulatory agency, which can be broken down so that each manufacturer pays an appropriate share. Total user fees should not be formulated so that they decrease if the prevalence of product use decreases. Instead, if prevalence decreases, user fees per product sold should increase. This will serve an additional purpose in that per product prices will increase if prevalence decreases.

It is very important that the user fee structure does not oblige a regulatory agency to allow marketing or encourage an increase in tobacco product sales. For example, user fees should not be based on the total number of products sold, with user fees increasing as prevalence increases. This could cause a conflict of interest in the regulatory agency. The design of the user fee system should encourage reductions in tobacco product sales or, at a minimum, have no impact. It must not be designed so that it encourages an increase in overall tobacco product sales. When regulatory agencies evaluate whether to allow the marketing of products, the payment of a user fee based on this decision must be the same whether the decision is to allow or deny marketing. User fees must not be based on a positive decision to allow marketing of a product.

Fines can provide another source of revenue for regulatory activities, although it should not be the sole means of funding. Examples include monetary penalties for unlawful or non-compliant activity, including failure to report. This approach encourages compliance by manufacturers. A set fee for business activity not related to proportion of sales is another possible source of revenue to cover costs.

When using any of these funding mechanisms, a firewall should be set up between the manufacturer and the agency carrying out regulatory activities to prevent the manufacturer having undue influence. This can be accomplished by requiring the manufacturer to pay the appropriate funds

Chapter 1   Testing in the Context of a Country's Regulatory Authority

into the national treasury, with the regulatory agency receiving a suitable appropriation. But national governments are often looking for sources of funds to support a myriad of activities, so any mechanism must be clearly described by law and ensure the continuous, certain and appropriate funding of regulatory activities.

1.8   Implementation

An important question when considering how to create a tobacco product testing programme is whether to implement everything at once, or step-by-step. When possible, a step-by-step approach is generally recommended. This allows the programme to start more quickly, since incremental steps can be taken to address obvious needs instead of having to anticipate every future possibility. In addition, it allows regulatory agencies to learn from initial mistakes and make adjustments. If an all-at-once approach is taken, it may be so burdensome to change direction that initial decisions may hinder the programme into the future.

Under certain circumstances, the situation may require an all-at-once approach. If the political conditions are such that a gradual approach is not possible, regulatory agencies may be required to start immediately. While this may be possible, there are significant hazards to this approach. Because of the possible pitfalls that agencies may encounter, there is a higher need for careful consultation with experts both before and during the process of creating the testing programme. Details of the various approaches to establishing laboratory capabilities are given in the next four chapters. See section 5.2 for a discussion of expertise available from TobLabNet.

Chapter 2   Introduction to Three Possible Routes to a Testing Laboratory

Operating a laboratory and maintaining the necessary quality of laboratory measurements can be costly and resource intensive. Tobacco product testing laboratories require experienced staff who have successfully carried out analytical measurements that will be heavily scrutinized and challenged. Testing laboratories require effective laboratory information management systems that can efficiently process, evaluate and store large amounts of data. These requirements differ from the requirements of typical research laboratories because of the nature of the work and the intensity of the workload. Laboratories require expensive analytical equipment that must be maintained, serviced and replaced periodically. Modern laboratory instruments are very complex devices that require particular expertise to ensure they operate properly and meet specifications.

Laboratories must maintain day-to-day analytical reliability so that all results are consistent and accurate. Laboratories require external accreditation and quality monitoring to demonstrate the quality of results, their dependability and to demonstrate their fidelity when under the intense scrutiny to which they will be subjected. To accomplish all this, there must be a guarantee of regular and sufficient support of both funding and personnel resources for any laboratory to maintain its testing capability. Competence is developed over time and must be maintained so that it can be relied upon when needed. We suggest three approaches

Chapter 2 Introduction to Three Possible Routes to a Testing Laboratory

to creating laboratory capacity for the testing of tobacco products design, contents and emissions.

This chapter summarizes these approaches, and subsequent chapters provide further detailed information on this capacity.

2.1 Contracting with an external laboratory

There are several experienced, independent tobacco testing laboratories around the world that are not affiliated with the tobacco industry. TobLabNet was set up to encourage the development of such laboratories and to better assure the quality and consistency of measurements. Laboratories, such as these, generally take two forms: independent commercial tobacco product testing laboratories and government-owned/operated tobacco product testing laboratories.

If they have been in operation for a substantial period of time, independent commercial tobacco product testing laboratories have certain advantages. They should already have capabilities that have been adequately tested and verified, and have experience participating in inter-laboratory comparisons. They already have experienced scientists and technicians on staff and the equipment needed to carry out a range of analyses. They will be accustomed to testing and reporting on a contractual basis and prepared to provide results under those conditions. They will already have developed IT systems and should already participate in an external quality assurance programme. For a regulatory agency ready to have testing done, these laboratories can quickly respond and provide results in a timely manner. But in general, they are limited to their own menu of testing capabilities. It may be possible for them to develop new capabilities, but

TOBACCO PRODUCT REGULATION
Building laboratory testing capacity

this would take time and they would need assurances that developing and validating a new method would be commercially beneficial. So a country-specific test may not be an immediate priority.

Some independent commercial tobacco product testing laboratories also make measurements on a contractual basis for the tobacco industry. This may concern some countries regarding their adherence to Article 5.3 of the WHO FCTC. In these cases, countries should require assurances of a firewall[3] to protect public health from commercial interests and to ensure confidentiality and independence of results. Laboratories that perform product testing for both the industry and regulatory agencies should not be automatically rejected, but evaluated to ensure their integrity, lack of bias and confidentiality.

There are also a substantial number of government-owned and operated tobacco product testing laboratories. TobLabNet continues to successfully work with several government-owned tobacco testing laboratories around the world to encourage the development of capabilities and provide a mechanism for inter-laboratory validation (see Appendix 1). Thus there are very effective and reliable government laboratories that understand the importance and objectives of regulatory testing of tobacco products and that face many similar challenges to those encountered in starting a new programme. These laboratories have many of the same advantages as working with independent commercial tobacco product testing laboratories, including experience, IT systems, quality assurance programmes and established capabilities. In addition, working with another country's regulatory agency can be a big advantage for a country

---

3 A system to ensure that data and information are protected so that public health and commercial interests are separate and not accessible to each other.

Chapter 2 Introduction to Three Possible Routes to a Testing Laboratory

that is just starting a programme.

For example, this interaction can provide a natural consultation relationship between new and experienced programmes. If a new regulatory agency is making unsuitable decisions (e.g., testing of the incorrect analytes in emissions), government agencies are more likely to provide advice in place of carrying out an inappropriate measurement. This could be a major advantage for a new tobacco regulatory programme. The biggest disadvantage of working with an established government laboratory is that they have their own statutory requirements and priorities. So they are not as likely to be available to carry out measurements in the time frame desired. They may be delayed by other priorities and their management is likely to require that their own priorities take precedent.

The third option for external laboratories is those owned and operated by the tobacco product manufacturers themselves. These should be avoided under any circumstances since there is an inherent conflict of interest.

2.2 Using an existing internal laboratory

For this discussion, we are assuming that a country has an experienced laboratory that is already doing testing for other purposes. For example, the country may conduct environmental or pharmaceutical testing and has previous experience of compliance testing and reporting. There are certain advantages and disadvantages to this approach.

For a laboratory already testing other consumer products, the development of internal capabilities to test tobacco products will have a foundation upon which to build. One of the biggest challenges in developing an effective laboratory where none existed before is hiring of staff with

TOBACCO PRODUCT REGULATION
Building laboratory testing capacity

valuable expertise who understand how to carry out valid and legally-defensible measurements. In addition, much of the laboratory equipment, IT systems and quality assurance programmes will already be in place and can be adopted for tobacco product testing purposes. Such an approach will be cheaper and quicker than creating a tobacco-testing laboratory from scratch. There may also be other advantages to using this approach. If funding is inadequate for the current laboratory to be as effective as desired, additional funds from tobacco testing could help. This would be a major advantage for laboratories which are often provided with limited government funding.

On the negative side, tobacco testing is likely to require new equipment and expertise. For example, smoking machines and expertise in their use are limited to combusted tobacco product testing; environmental or pharmaceutical testing laboratories will not have this equipment or experience. Thus, acquiring this capability will still require some significant start-up time and costs. But if planned correctly, this could be a second stage in laboratory development if there are specific priorities (e.g., cigarette tobacco content or smokeless tobacco testing) that do not require this capability. In addition, as with the situation described above, developing tobacco product testing capabilities in an existing laboratory may result in priority conflicts. For example, a drug testing laboratory is likely to already have fully assigned staff and equipment. Rarely do laboratories have significant excess capacity. So a natural conflict will occur at times when both programmes need results quickly. It would be wise to specify clearly how such conflicts will be addressed before final agreement is made.

If there are research laboratories available that may add tobacco product

Chapter 2   Introduction to Three Possible Routes to a Testing Laboratory

testing capabilities, be aware that the nature of the work and the scientific approach are not the same. The work of a laboratory that tests products for compliance or reporting purposes has requirements that are different from those of a research laboratory. In general, research laboratories would need to increase IT infrastructure, put more robust quality assurance systems in place, seek accreditation, and be prepared to provide forensic evidence in order to be successful as a compliance laboratory. By contrast, an existing internal compliance laboratory should already be accustomed to generating results that can be used for compliance or legal purposes. A Case Study with a pre-existing tobacco testing laboratory using other facilities to support additional tobacco product testing capabilities is described in Section 4.3.

2.3   Developing a dedicated laboratory

The final option to be considered is developing a dedicated government tobacco product testing laboratory without sharing resources. This approach has some considerable advantages. Having a dedicated testing laboratory means that the priorities of testing capacity and developing new testing capability will be driven by tobacco regulation priorities. Thus the priorities for use of available capacity can be set by a single management structure. It is also possible to use any excess capacity as a means of generating additional revenue to support the laboratory's operations.

On the other hand, developing a laboratory that can generate completely reliable results will require significant commitments of time, funding and human resources. This may be alleviated somewhat if there are current laboratory facilities or even facilities and staff that can be reassigned to a new mission. If not, this could require construction or remodelling of physical structures. Laboratories require special air handling, power requirements (such as uninterruptible power supplies) and other physical facilities that are not typically present in office, retail or commercial buildings. This may mean building new facilities or conversion of current facilities. It may be challenging to maintain adequate support to develop a laboratory that requires years of construction and outfitting, especially when government has other budgetary priorities. Maintaining support for testing capabilities among government decision-makers is likely to require data to demonstrate the value of this significant investment; delays could result in loss of support. While this is certainly a viable option, several countries have been unsuccessful when trying this approach. Their lack of success has been largely the result of delays in construction and changes in government priorities as administrations change.

This advantages and disadvantages described above are summarized in Table 1 and described in more detail in the chapters that follow.

Table 1. Advantages and disadvantages of approaches to developing laboratory capability

Laboratory type	Pros	Cons
External laboratory – commercial	• Internal measurement expertise not required • Lower start-up costs • Faster start-up than developing new capabilities • Broad capabilities • Immediate access to experienced scientists • Recognized validity of results	• Availability not guaranteed • Reliability must be regularly assessed • May be more expensive long-term • May be limited to methods available in that laboratory • Less flexibility to develop new methods

Chapter 2 Introduction to Three Possible Routes to a Testing Laboratory

Continued

Laboratory type	Pros	Cons
External laboratory – government	• Developing new internal expertise may not be required • Less expensive start-up costs • Faster start-up than developing new capabilities • Immediate access to experienced scientists • Recognized validity of results • Encourages consultation with regulatory agencies from other countries	• Possible delays in generating results • Reliability must be regularly assessed • May be more expensive long-term • Less flexibility to develop new methods • May have limits to capability and capacity
existing internal laboratory	• Some expertise available • Improved overall efficiency through resource sharing • Lower start-up costs • Faster start-up than creating a new lab • Pre-existing and reliable IT and quality systems • May help stabilize funding for both programmes	• Availability depends largely on other priorities • Must purchase some tobacco-specific equipment (e.g. smoking machine) • Must develop some expertise (e.g. smoking machine operation) • Will require some start-up time and costs
Dedicated internal laboratory	• Guaranteed availability • Can generate revenue through outside work • Can develop capability as needed • Complete flexibility of priorities	• Broad capability is expensive • Large start-up costs • May require building facilities • Will require significant start-up time • Must obtain expertise • Government support and funding level may fluctuate

For countries beginning a testing programme, starting with an experienced external laboratory is recommended (See Fig. 1). This serves several purposes, of which the three most important are listed below:

1. It allows the testing to be started quickly while support is strong.
2. It provides data quickly to help support the rationale for carrying out tobacco product testing long-term.

TOBACCO PRODUCT REGULATION
Building laboratory testing capacity

3. It lifts the burden of laboratory development and data quality from regulators, allowing them to concentrate on other critical issues.

If there is a significant sample load and support remains strong, a regulatory agency may consider an agreement with another government-owned and operated testing laboratory within the country to carry out tobacco testing. If the external contract is maintained, this will allow a step-wise transition with no loss of capability, but expanded capacity. Finally, if sample throughput supports it, consideration may be given to build and outfit a dedicated national tobacco testing laboratory. But this should only be done if it is evident that there is an adequate volume of testing and administrative support continues to be strong. As discussed above, going directly from no capacity to a dedicated national tobacco product testing laboratory has not proven to be a viable approach.

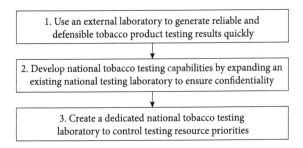

Fig. 1 Suggested approach to building testing capabilities

Chapter 3   Contracting with an External Testing Laboratory

3.1   Laboratory selection criteria

There are several important considerations when attempting to identify an external laboratory with which to contract. The emphasis given to each of these in making a final decision will vary from country to country, but all may be considered. It is worthwhile to investigate each of these considerations before making a final decision.

3.1.1   Experience in tobacco analyses

This is critical and possibly the most important factor. The primary reasons, as listed above, for working with an external laboratory are based on specific experience in performing tobacco product analyses. Choosing a laboratory with limited experience contradicts the advantages of using an external laboratory. As has already been described, tobacco analyses has its own requirements, procedures, equipment and standards, and a laboratory with years of experience in this field and a demonstrated track record will substantially reduce start-up costs and time.

3.1.2   Equipment capabilities

Analytical chemistry is constantly evolving to develop procedures and equipment that are more accurate, reproducible, selective and sensitive. While improved sensitivity is not necessary when analyte levels are high, many ingredients and emissions in tobacco products are at levels that challenge the ability to detect and accurately quantify. Modern analytical equipment is steadily improving detection and quantification limits such that these improvements may be critical for answering specific tobacco regulatory questions. Evaluation of an external laboratory should include an assessment of the breadth and sophistication of analytical instrumentation. In order to address immediate needs, laboratories should include, at a minimum, the equipment listed in Table 3. Countries should also consider which of the equipment listed in Table 4 and any additional equipment that may be needed for country-specific analyses. Any laboratory to be considered must include gas chromatography (GC)-flame ionization detection (FID) and liquid chromatography (LC)-diode array detection instrumentation. It is an advantage to also have LC-tandem mass spectrometry. In addition, the degree to which automated sample preparation apparatus are used will help reduce human error. It is expected that tobacco manufacturers will have access to the most advanced instrumentation available and when comparing data in a compliance, enforcement or legal setting, it is best to know that the instrumentation used is the most advanced for a particular purpose. Decisions should be based, wherever possible, on state-of-the-art analyses.

3.1 3 Staff qualifications

An experienced and capable staff is the most valuable commodity in a laboratory. While advanced equipment can be purchased, staff must be trained and experience requires time to develop. Experienced laboratory analysts and instrument operators are necessary for making reliable analytical measurements on any commodity, but especially so for tobacco products because of the particular requirements. When assessing a laboratory, an evaluation should be made of the number and length of time that staff members have been performing analytical measurements as a whole, and tobacco product analyses in particular. In addition, a system for regular training of new and experienced staff should ensure they are competent and evaluated before they start independent analyses and to make certain that they keep up with

advances in the field.

3.1.4 Breadth of capabilities

Even if the initial range of analyses is limited, it is likely that, at some point, the situation will change and there will be the need to perform new tests. A laboratory that only has limited capability may not be able to perform new analyses when needed for regulation. In that case they will require significant start-up time to develop capabilities and verify new measurements. If those capabilities are already present, this start-up time would be minimal. In addition, a laboratory with a wider breadth of capabilities is likely to be of higher quality. Capable scientists are always looking for new challenges or ways to improve. A laboratory that allows staff to grow in their jobs attracts and retains better-qualified scientists.

3.1.5 Excess capacity

In order for testing results to be useful for regulatory purposes, they must be accurate, reproducible, sensitive, and selective (see Section 4.3 below).

But, they also must be timely. If a laboratory is not able to provide results when needed, the critical opportunity may have passed, and these data may no longer be relevant or have the most impact. Capacity is not only the sample throughput when all systems are operating as expected, it also offers flexibility. For example, a tobacco smoke testing laboratory that has only one smoking machine cannot process any combusted product emission samples if it is being repaired. A facility that has duplicates of all critical equipment and backups for all staff members can continue to function when unexpected events occur. Capacity should be included in an overall laboratory assessment.

3.1.6 Proven track record

Advanced equipment, trained staff and excess capacity do not ensure that laboratories can produce and report reliable results within deadlines. Timeliness can also be a function of the institutional culture and the effectiveness of management. Laboratories should be able to provide references or records showing that reliable results can be regularly produced and reported on time. It is worthwhile, if possible, to obtain a list of previous customers and contact some of them randomly to evaluate the ability of the laboratory to produce results, as promised.

3.1.7 Accreditation

Accreditation is the documentation by an independent body showing a laboratory has the systems in place that should enable them to produce reliable results that have been adequately tracked and verified. There are international and national laboratory accreditation bodies and accreditation standards such as ISO 17025 that effectively carry out this function (see Section 4.2). Any laboratory generating results for compliance and enforcement purposes should be accredited. Any laboratory that is not should immediately be removed from consideration. Even so, it is important to keep in mind that accreditation alone does not guarantee valid results. It is possible for laboratories to be accredited and not be able to produce adequate results.

3.1.8 Ability to add new methodologies

As discussed above, there will be times when a specialized analysis is needed that was not previously anticipated and for which there are no current methods. A good example is the introduction of a new product type or product modification likely to generate new emissions. In that case, a laboratory may need to develop and validate a new method in a relatively short time. Laboratories should be able to give examples of carrying out this process from previous instances. Because this could require substantial development efforts, the costs for new methodology analyses will likely be higher than for a routine measurement.

3.1.9 Information technology systems

Accuracy is a fundamental requirement of analytical measurements. Accuracy can be maximized by using the right analytical methods and instrumentation carried out by trained and experienced staff. But errors can occur whenever the analytical process involves manipulation of data. This is a particular challenge whenever numbers are transcribed by hand. The less hand transcription of numbers, the fewer errors. Laboratory information management systems (LIMS) are common throughout the laboratory testing community and are considered a necessity for testing and reporting. A laboratory without a LIMS should be dropped from further consideration. The more automation, the lower the chance of human error, but there also needs to be a process for checking that the LIMS works properly. A LIMS that tracks samples and processes data from sample receipt to report preparation is highly desirable. But, checks also need to be built into the quality assurance programme that regularly validate these systems.

3.1.10 Quality assurance programme

Quality assurance combines a well-defined quality control programme with an overall mindset of quality. Quality control evaluates whether systems continuously operate within standard parameters. Quality assurance ensures that the systems were designed correctly and are operated appropriately. A quality assurance programme involves training as described above, management review of compliance with laboratory standards, and regular review of results before they are reported. An effective quality assurance programme is critical for data reliability.

3.1.11 Participation in inter-laboratory validation

It is very likely that any results generated will need to be compared to results from other sources. These may be results generated historically by other laboratories or in other countries. It is also important, from a legal standpoint, to show that data from a particular laboratory is comparable to data from other laboratories. Inter-laboratory validation exercises occur on an international basis and go by several names including round-robin and inter-laboratory comparison. In these exercises, the same samples are analysed by multiple laboratories and sometimes using multiple analytical methods. Results are compared to determine what is the consensus mean between all laboratories and the deviation of each laboratory from the mean. Participation in round-robin inter-laboratory validation can help address an area of uncertainty that could be critical in the use of the data for public health regulatory purposes. Another substantial advantage of participation in round-robins is that, if there are high priority analyses and a laboratory is fully occupied or instruments are not operational, other laboratories can be utilized with an assurance of data comparability.

3.1.12 Cost of analyses

Cost may seem to be a major consideration, but it is one of the least important in deciding which laboratory to use. If tobacco product manufacturers are paying the cost of analyses, this should not be a critical issue for the regulatory body but should be noted as part of the overall assessment.

3.1.13 Other customers

There may be a conflict of interest within the laboratory with analyses performed for other customers, such as tobacco industry clients. As discussed, above in section 2.1, there are ways to address these concerns, but they should be considered as a factor when making a final decision.

3.1.14 Location

The proximity of the laboratory to the country requesting analyses is primarily a question of logistics, but can also be a matter of import law. The most critical factor is whether there are reliable means of transporting samples from the site of collection to the laboratory efficiently. Since samples will most likely be shipped through a common carrier, this is typically not a major barrier, but the process and length of time required should be evaluated beforehand. Samples that are not adequately stored during shipping may be altered and that could raise questions about data integrity.

Some countries have restrictions on tobacco imports and transferring such products across national borders can be problematic. But there can be allowances for products sent for testing purposes only. It is advised that this issue be clarified before deciding to use a laboratory in another country. It should also be considered that a laboratory that is close in proximity may be easier to visit for inspection purposes than one located far away.

The form below (See Fig. 2) provides a convenient means to organize a laboratory rating. The factor under weighting should be adjusted based on the specific country's requirements. A possible set of weights is provided, but should be adjusted as appropriate. These weights were based on factors which the author believes would enable a laboratory to be most successful in carrying out accurate testing of a wide range of analytes in tobacco products. The score should be determined for each laboratory. Then the product (weighting multiplied by the score) calculated and the sum added

Chapter 3 Contracting with an External Testing Laboratory

on the bottom line.

Laboratory Name _____

Factor	Weighting (1-10)	Score (1-10)	Product (weighting x score)
Experience in tobacco analysis	10		
Equipment capabilities	8		
Staff qualifications	8		
Breadth of capabilities	6		
Excess capacity	6		
Proven track record	8		
Accreditation	8		
Ability to add new methodologies	6		
Information technology systems	6		
Quality assurance programme	8		
Participation in inter-laboratory validation	8		
Cost of analyses	2		
Other customers	4		
Location	4		
Total (sum of above)			

Fig. 2 Laboratory Rating Sheet

3.2 WHO TobLabNet

The WHO Tobacco Laboratory Network (TobLabNet) is a network of government, academic and independent laboratories designed to strengthen national and regional capacity in the testing of tobacco product contents and emissions. *(20)* In April 2005, the WHO Tobacco Free Initiative (TFI) established WHO TobLabNet based on the aims and objectives of Articles 9 and 10 of the WHO FCTC and the

TOBACCO PRODUCT REGULATION
Building laboratory testing capacity

recommendations of the WHO Study Group on Tobacco Product Regulation (TobReg). TobLabNet is a primary source of laboratory support, methods development, and scientific information in the areas of tobacco testing and research for national governments to fulfil their requirements and needs related to the WHO FCTC.

Originally, there were 25 laboratories from 20 countries representing all six WHO regions who agreed to be a part of WHO TobLabNet. Over the years, the participation of laboratories in method validations has varied depending on national priorities and availability of resources. The current list of participating laboratories is given in Table 2.

The goal of WHO TobLabNet is "to establish global tobacco testing and research capacity to test tobacco products for regulatory compliance, to research and develop harmonized standards for contents and emissions testing, to share tobacco research and testing standards and results, to inform risk assessment activities related to the use of tobacco products, and to develop harmonized reporting of such results so that data can be transformed into meaningful trend information that can be compared across countries and over time". *(21)*

To accomplish this, laboratories work together and support each other in collaborative projects guided by various lead laboratories. WHO TobLabNet works actively to provide advice to national governments seeking to develop and improve tobacco testing laboratories as a means of increasing capacity and ensuring consistency.

WHO TobLabNet carries out work requested by the Conference of the Parties to the WHO FCTC through the WHO FCTC Secretariat, under the auspices of WHO, for accomplishing objectives set out under the WHO FCTC. Recently, this work has involved method development, validation

Chapter 3 Contracting with an External Testing Laboratory

and verification for measuring high priority contents and emissions in commercial cigarettes and other tobacco products. In addition, round-robin testing of the methods by various laboratories have been used to measure the inter-laboratory reproducibility of these methods. The current list of constituents and their status is found in Table 3.

Governments seeking information on establishing mechanisms for tobacco products testing are advised to contact WHO TobLabNet for advice and guidance. Based on the availability of resources, WHO TobLabNet may be able to provide training and support capacity building of laboratories looking to begin tobacco testing or expand current capabilities, both within the network itself and for laboratories looking to become a part of the network in the future.

Table 2 List of Tobacco Laboratory Network (TobLabNet) member laboratories

WHO Region	Country	Laboratory
Regional Office for Africa (AfRO)	Burkina faso	Laboratoire National de Santé Publique
Regional Office for the Americas (AMRO)	Canada	Labstat International ULC
	Costa Rica	Instituto Constaricence de Investigación y Enseñanza en Nutrición y Salud (INCIENSA)
	Mexico	National Institute of Public Health
	United states of America	Centers for Disease Control and Prevention
		Alcohol and Tobacco Tax and Trade Bureau (TBB)
		Battelle Public Health Center for Tobacco Research
		Virginia Commonwealth University
		National Cancer Institute
Regional Office for South-East Asia (SEARO)	India	Directorate general of Health services
	Indonesia	National Agency of Drug and Food Control

TOBACCO PRODUCT REGULATION
Building laboratory testing capacity

Continued

WHO Region	Country	Laboratory
Regional Office for Europe (EURO)	Albania	Institute of Public Health
	Bulgaria	Tobacco and Tobacco Products Institute
	finland	National supervisory Authority for Welfare and Health
	france	Laboratoire National de Métrologie et d'essais
	Germany	Federal Institute for Risk Assessment (BfR)
	Greece	General Chemical State Laboratory of Greece
	Ireland	State Laboratory
	Italy	European Commission, Joint Research Centre
	Lithuania	National Public Health Surveillance Laboratory
	netherlands	Laboratory for Health Protection Research of the Dutch National Institute for Public Health and the environment
	Russian federation	All-Russia Research Institute of Tobacco, Makhorka and Tobacco Products
	spain	Agrarian and food Laboratory
	Switzerland	L'Institut universitaire romand de santé au travail (IST) Lausanne
	Ukraine	L.I. Medved's Research Center of Preventive toxicology
Regional Office for the Eastern Mediterranean (EMRO)	Lebanon	American University of Beirut
	United Arab emirates	National Laboratory & Research Center
Regional Office for the Western Pacific (WPRO)	China	China Centers for Disease and Control and Prevention
		Institute of Tobacco Safety and Control
	Japan	National Institute of Public Health
	Republic of Korea	Ministry of food and Drug Safety
		Korea Centers for Disease Control and Prevention
	singapore	Health sciences Authority
	viet nam	National Institute of Occupational and Environmental Health

Table 3 Current TobLabNet Method Development Status

Method	Analytes	Matrix	Analytical Method	Status
Nicotine	Nicotine	Tobacco	GC/FID[a]	Validated
Ammonia	Ammonia	Tobacco	Ion chromatography/ conductivity detection	Validated
Humectants	Propylene glycol Glycerol Triethylene glycol	Tobacco	GC/fID (GC/MS)[b]	Validated
TNCO	Tar, nicotine, carbon monoxide	Smoke	GC-FID (nicotine) GC-TCD[c] (water for tar calculation) Non-dispersive infrared analyzer (for CO)	Validated
TSNAs	N-Nitrosonornicotine (NNN) 4-(Methylnitrosamino)-1-(3-pyridyl)-1-butanone (NNk) N-Nitrosoanatabine (NAT) N-Nitrosoanabasine (NAB)	Smoke	HPLC/MS-MS[d]	Validated
BaP	BaP	Smoke	GC/Ms	Validated
VOCs	Benzene 1,3-Butadiene	Smoke	GC/Ms	Validated
Carbonyls	Formaldehyde Acetaldehyde Acrylaldehyde	Smoke	HPLC DAD[e]	Validated

[a] Gas chromatography/Flame ionization Detection
[b] Gas chromatography/Mass spectrometry
[c] Gas chromatography/Thermal; conductivity detector
[d] High-performance liquid chromatography/Tandem mass spectrometry
[e] High-performance liquid chromatography/Diode Array Detector

3.3 Agreements and legal/ethical issues

As stated above, national regulatory agencies should consider their specific legal and ethical issues when planning to contract with an external laboratory. These should be understood and resolved before an agreement is signed and before analyses are performed. It is very important that the data obtained by regulatory agencies be adequate for their purposes; all requirements related to data quality for the use of the data should be clearly described in any agreement.

The first consideration when entering into an agreement with an external laboratory is ensuring the laboratory's independence from the tobacco industry. Article 5.3 of the WHO FCTC urges Parties to protect tobacco control policies from "commercial and other vested interests of the tobacco industry". WHO TobLabNet applies strict membership criteria which excludes laboratories that are totally or partially owned by a tobacco company, or laboratories with persons in senior management positions employed by or affiliated with the tobacco industry. Laboratories that receive funds from the tobacco industry must additionally demonstrate independence. These conflict-of-interest requirements ensure that public health objectives of testing policies are never compromised.

Another consideration is the legal requirement for introducing testing data as evidence in legal proceedings or as the basis for regulatory action. For example, it should be understood from the start that the laboratory has the required accreditation and evidence of data quality that are needed to use the data in taking action. Chain of custody may be a critical element in assuring that the data is from analyses of the materials intended to be tested. An assurance of sample integrity can be a critical factor in the legal acceptance of data.

Certain laboratories may consider background information related to the analyses that are performed as proprietary. Detailed descriptions of laboratory methods, quality control results, findings from method validation, or information on the results of round-robins could be

Chapter 3 Contracting with an External Testing Laboratory

considered private and not for release. If this information is needed as part of regulatory evidence, it should be clear from the outset that the laboratory must make this information available. Discovering otherwise, after the analyses have been completed, would greatly restrict the effective use of the data.

Another pre-contract consideration is the reporting of analysis results. Laboratory measurements typically involve multiple replicate analyses on each sample. In other words, in order to improve accuracy, account for the variability in a commercial product made from an agricultural product such as tobacco, and to evaluate reproducibility, multiple measurements are made and these are averaged to arrive at the final result. Typically, this may include anywhere from three to 20 replicates for a single final result. The number of replicate analyses to be made is a balance between sample availability, cost (more replicates cost more) and data quality (more replicates reduce random variability).[4] The regulatory agency should consider how best to balance the requirements of the data use versus the cost of the analyses. When choosing laboratories, the number of replicates typically carried out should be part of the overall consideration.

In the report provided to the customer, laboratories can report all of the replicates or can report the final average. Regulatory agencies should require the reporting of all replicates so that the data quality can be fully assessed and the appropriate statistical analyses can be carried out as suitable for the particular application. This should be included in the agreement with the laboratory so that the required information is retained and reported.

---

4   More detail on this subject can be found in Appendix 1.

TOBACCO PRODUCT REGULATION
Building laboratory testing capacity

When using an external laboratory, it is important to understand if and when those data are available to parties other than those requesting and funding the analyses. Laboratories should have adequate firewalls so that business agreements with other involved parties including tobacco product manufacturers do not influence the analyses for the regulatory agency. Regulatory agencies should investigate whether laboratories have a potential conflict of interest before agreeing to use them. In addition, regulatory agencies should be aware of public access or freedom of information laws, which might impact access to the data by interested parties. These laws may not influence whether a regulatory agency carries out testing using a particular external laboratory, but they should be aware of issues that could arise.

Even when it is understood that the analyses will be funded by the industry, there are at least three options for the reporting of results: government only, government and industry simultaneously, or industry only, which then provides data to the government. From a reliability standpoint, reporting results from the laboratory straight to the government is preferable if that accords with the requirements defined in the statutes or regulations that require these reports. This is possible either through reporting results to the regulatory agency alone, or simultaneously to the regulatory agency and the manufacturer whose products are being evaluated. Reporting directly to the government agency helps to reinforce to the laboratory that the regulatory agency is the end user of the results and a decision-maker concerning where analyses may be performed. Simultaneous reporting is the most likely acceptable option for all parties involved because of lack of trust. But reporting by the laboratory to the manufacturer, which then provides these data to the regulatory body, should be avoided since that provides an opportunity for data manipulation and creates unnecessary uncertainty.

3.4 Sample load estimates

Laboratory capacity is a significant consideration when choosing a laboratory because timely results are critical for providing data to decision-makers when issues are ripe for action. But excess laboratory capacity does come with a cost. For a laboratory to be prepared for a large sample load or unexpected requests, they must have access to excess equipment and staff. These resources must be supported even when not in use, so overhead costs increase. It is very important that regulatory agencies provide good estimates of anticipated sample load so that laboratories can be prepared ahead of time to provide a timely response. Poor estimates of sample load or timing of sample delivery causes delays in results reporting or wasted resources, the cost of which must be passed on to the customers.

In order to best optimize available resources, sampling and analysis requests should be spread over the entire year and not just at one time of year. There are several efficient ways of doing this. Manufacturers can be required to submit samples at different times of the year, perhaps spreading this out over all four quarters. Or manufacturers could submit one quarter of their brands/subbrands for analyses each quarter of the year. This spreads the laboratory workload, allows a more efficient use of resources and helps to control the cost of each analysis.

When choosing laboratories, it is important that evaluation is based on capacity and the ability to meet specified time frames. Some consideration must be made for unforeseen circumstances, and this should be considered when estimating sample reporting expectations. Regulatory agencies need to consider how soon results will be required after submission and whether there may be a need for special priority requests. This should be included in an agreement when contracting with a laboratory.

3.5 Costs

The cost for tobacco product contents and emissions analyses can be high relative to other regulatory activities. No matter whether manufacturers submit samples directly to the laboratory or if the samples are first submitted to the regulatory agency and then submitted to the laboratory, all costs must be covered by the tobacco product manufacturers. If a manufacturer believes that the cost of testing is excessive, they can choose not to market their products or reduce the number of brands/subbrands sold in a particular market. In either case, a reduction in the number of brands/subbrands sold in a country could result in a reduction in overall tobacco product use and an improvement in public health.

Depending on the analysis to be performed, the matrix (tobacco or smoke), the equipment used, and the number of replicates, the cost of each individual analytical test in each individual brand/subbrand can range from hundreds to thousands of US dollars. Measurements in smoke are generally more expensive than measurements in tobacco because the process of generating and collecting smoke using an appropriate regimen adds an additional step to the analytical process. Some analyses can be performed using a variety of analytical equipment. For example, analysis of benzo[a]pyrene in mainstream smoke may be made using infrared spectroscopy, gas chromatography with mass spectrometry, high-performance liquid-chromatography (HPLC) with fluorescence detection, HPLC with mass spectrometry (MS), or HPLC with tandem MS. When a lab can perform adequate cross-method validation, use of alternative methods with other analytical instrumentation may be possible. The cost of purchasing and maintaining this equipment varies so that the cost of the analysis will be different. There may be certain critical benefits (sensitivity and selectivity) to using more sophisticated and sensitive analytical equipment, but the use of higher cost equipment when not needed should be avoided. Analytical measurements should be evaluated for their fitness for use before agreeing to carry out a higher cost analysis.

In certain cases, multiple constituents can be measured simultaneously using the same analytical method. For example, most analytical methods for TSNAs measure all four of these compounds using the same method. The cost is not substantially lower if only one of the TSNAs is measured. The case is the same for some PAHs, heavy metals, aldehydes, and many volatile organic compounds. So the overall cost of testing and reporting is less dependent on the number of constituents reported, but rather the number of analytical methods that must be performed to make the measurements.

3.6 Case study – Canada

Canada provides a good example of a country that has successfully used an external laboratory for research purposes as well as method development in support of tobacco control. In the late 1970s, Health Canada started contracting the services of an independent laboratory[5] (not affiliated

---

5 Labstat in Kitchener, Ontario, Canada (http://www.labstat.com/servicesoverview.html).

with the tobacco industry) to test the contents and emissions of various tobacco products on the Canadian market. In the 1980s, after having identified the need to measure a larger selection of analytes in the contents and emissions of tobacco products, Health Canada called upon the same independent laboratory to develop laboratory methods, work which continued throughout the 1990s and led to methods covering 20 analytes found in the contents and 40 in the emissions of tobacco products. *(22)*

The *Tobacco Reporting Regulations*, which came into force in 2000, incorporate these as "Official Methods", which manufacturers must use for reporting to Health Canada. *(23)* As per section 4 of the regulations *(22)*, tobacco product manufacturers must report results using a laboratory that is accredited under the International Organization for Standardization standard ISO/IEC 17025, entitled *General Requirements for the Competence of Testing and Calibration Laboratories*. Typically, independent laboratories are retained by the manufacturers to meet their reporting obligations. Health Canada continues to use independent laboratories as needed to support analytical development projects.

The use of a single analytical laboratory with a defined set of analytical methods has a clear advantage. When measurements are made by the same laboratory using the same methods, uncertainty is significantly reduced and there is increased confidence in the direct comparability of all results. The disadvantage to this approach is the dependence on a single laboratory. For example, if that laboratory ceased operations, there would be challenges in moving the necessary analyses to a different laboratory. Also, there may be concerns about sample analysis capacity if only one laboratory is used. If only a single laboratory is identified, it would be prudent to have the laboratory provide a backup plan describing what steps could be taken to mitigate any loss of capacity and ensure continuity of operations.

3.7 Step-by-step process

1. **Do your own internal research.**
    a. Identify a reasonable first set of analyses for your tobacco testing that would address your country's priorities.
    b. Identify the analytical requirements for the intended use of the data.
    c. Determine the estimated initial workload (number of samples over what period). This may be determined by the number of brands/subbrands being marketed and the frequency and schedule of testing.
2. **Discuss your approach with another country's regulatory agency that has experience with tobacco product testing.**
3. **Evaluate available laboratories.**
    a. Check the requirements listed above and rank available laboratories.
    b. Evaluate the ability of laboratories to meet analytical accuracy, reproducibility, sensitivity and selectivity requirements to identify satisfactory laboratories.
    c. Use estimates of capacity requirements to identify acceptable laboratories.
4. **Discuss particular needs and requirements for expected sample workload, turnaround time, reporting requirements, etc. with the identified laboratory.**
5. **Finalize any required contractual agreement.**
6. **Communicate to the manufacturers, if appropriate.**
    a. what needs to be tested and when?
    b. which labs are acceptable?

Chapter 4 Using an Existing Internal Testing Laboratory

In certain cases, a national government may already have laboratories to test non-tobacco consumer goods. These may be test purity or to verify levels of therapeutic drugs in pharmaceuticals, cosmetics or imported goods. There may also be laboratories to test environmental samples such as air or water. The existence of government testing laboratories for other consumer goods or environmental samples provides an opportunity to readily develop government tobacco testing capabilities.

A tobacco regulatory agency may find that government organizations that already operate testing laboratories are willing to collaborate to expand their capabilities and include some tobacco design, content and emissions testing as part of their analytical portfolio. Because they are likely to be already overworked and under-resourced, this may require some careful persuasion by highlighting the advantages that this could bring to their activities. Explaining how expanding to include tobacco would benefit their current programme could help to convince them.

There are several advantages to working together with current government testing laboratories, when possible. The existing laboratory is likely to have equipment, supplies, trained personnel and quality assurance and information systems in place. Equipment alone is expensive to obtain and maintain when establishing a laboratory. Analytical equipment can cost as much as hundreds of thousands of US dollars per item and maintenance and repair is an ongoing requirement. If the current laboratory already

Chapter 4 Using an Existing Internal Testing Laboratory

has the equipment in place and has budgeted funds for repair and maintenance, those costs will be already covered, and additional analyses for tobacco products will not increase these costs substantially.

Laboratory personnel are a critical asset. It can take years to get the training and experience to develop an effective analyst. While it may be possible to hire experienced analysts directly, this can be a significant challenge because of their limited availability. If a laboratory already has experienced analysts, as would be the case for an existing testing laboratory, the time required to train them to carry out methods for analysis of tobacco products will be significantly less. If it is a priority to generate results quickly to illustrate the value of tobacco product testing, collaborating with an existing laboratory is a much better approach than creating a laboratory from the ground up. The value of experienced analysts cannot be overemphasized. The knowledge and experience of staff carrying out the analyses is the most critical factor in determining whether reliable results can be produced efficiently.

Similarly, the use of an existing laboratory allows the testing programme to start in a restricted way, if needed. When building a laboratory from the ground up, considerations must be made about the size of the laboratory and projected resources many years in the future. Thus initial design considerations must allow for the expected increases in capabilities and capacity. Invariably, this leads to overbuilding at the start of the programme with significant unused space. This may be challenging for a programme that is just beginning. By using existing laboratory facilities, tobacco testing can start at a limited level and then grow in a controlled manner as the programme expands.

Another significant advantage to this approach is that it can provide an additional source of funding for an existing laboratory. Laboratories invariably have to balance their instrument capacity and the number of staff with the expected workload and funding. Periodic variations in sample load are always a challenge for laboratory management. By expanding the sources and amount of funds flowing to the laboratory, variability in sample load can be better balanced. This is an important advantage that can be presented when discussing this approach with the current laboratory management. While there will be some increase in resource requirements, additional funding streams could help to address the highs and lows of sample analysis requests and funding.

In addition to smoothing out the laboratory funding stream, additional funding can support the purchase of additional equipment, hiring of new staff and expansion of other capabilities that would not be otherwise possible. Most of the analytical equipment for use in analyses of tobacco product contents and emissions can be used for other work. So if allowed, tobacco funding may be used to upgrade analytical instruments, which can then also be used for analyses of environmental samples or pharmaceuticals and other consumer goods. This would provide a significant advantage to the current laboratory. To the degree that synergism is possible, both programmes will benefit.

There are some disadvantages to this approach. Because of their current programmatic responsibilities, the current programme will likely view tobacco analyses as a lower priority, at least at first. It is imperative that priorities be discussed and an agreement put in place that clarifies how differences are to be resolved and how the two programmes will work together to meet all requirements. Otherwise, this could be a significant issue.

Chapter 4   Using an Existing Internal Testing Laboratory

In addition, there will be some analytical requirements for tobacco products testing that have no counterpart in analysis used in other testing programmes. The most obvious example of this is the need for the equipment and controlled temperature and humidity facilities required for the smoking of combusted conventional tobacco products. Existing laboratory management may hesitate to acquire these new capabilities. This should be discussed beforehand and an estimated timeline and plan clarified to prevent misunderstandings in the future.

4.1   Requirements (laboratory equipment, staff, overall cost) – how to identify the right laboratory

Requirements will be directly dependent on the initial analyses identified as the highest priority. If these analyses cannot be performed with available equipment, the benefits of using an existing testing laboratory will be limited. But much of the analytical equipment that is listed (in Table 3 above) should be available in a typical analytical laboratory.

A first step when assessing existing laboratories to which it may be appropriate to add tobacco analysis capabilities is to do a comparative analysis of available analytical equipment. A list of equipment that may be commonly found in a testing laboratory and used for analyses of tobacco product design, contents and emissions is given in Table 4. This list is broader than that presented in Table 3 and there are some duplications in this table because some analytes can be measured by more than one analytical instrument.

Previously, WHO identified useful equipment for developing a tobacco testing laboratory. *(24)* This list can be used when evaluating the instrumentation in an existing government laboratory to assess which existing laboratory might be best equipped and which equipment may still need to be purchased for particular tobacco testing applications depending on a country's priorities.

There is some additional equipment that will likely need to be purchased in order to carry out tobacco-product-specific analyses. This equipment should be considered in the overall plans for laboratory development.

- Smoking machine (w/non-dispersive infrared analyzer for carbon monoxide (CO)), approximately US$ 200 000
- Environmental chamber, approximately US$ 30 000

Section 3.1 above describes criteria for choosing an external testing laboratory. Except for experience in tobacco product testing, these criteria also apply when evaluating an internal laboratory that tests other regulated materials. The rating spreadsheet (Fig. 2) can also be used to identify those aspects that will need to be considered in this case. In particular, and in addition to the equipment available, when assessing the suitability of an existing testing laboratory, the following should also be evaluated.

- Adequate bench space.
- Effective information technology systems that can be used to track samples, limit data transcription and efficiently report and archive results.
- A quality assurance programme that meets accreditation requirements.
- Environmental control (temperature and humidity) that meets both current and anticipated instrument requirements.
- Adequate electrical systems that meet expected requirements by

Chapter 4 Using an Existing Internal Testing Laboratory

instrument manufacturers and ensure satisfactory instrument uptime.
- Well-trained, experienced staff who have a documented history of producing reliable analytical testing results.

Table 4 Analytical equipment for a tobacco testing laboratory

Instrumentation	Purpose	Approximate cost (US$)
Freezer(s)	Storage of samples	1,000
Analytical balance	Weighing of samples, "tar"	10,000
Pressure drop apparatus	Cigarette ventilation	40,000
Ion chromatography/ conductivity detection	Ammonia in tobacco	50,000
Continuous flow colorimetric analysis	Hydrogen cyanide in smoke	60,000
Chemiluminescence nitrogen oxide analyser	Nitrogen oxides	50,000
GC/FID	Nicotine in tobacco/smoke	100,000
GC/Thermal energy analysis	TSNAs in tobacco/smoke	150,000
HPLC/UV detection	Carbonyls in smoke	100,000
HPLC/Fluorescence	BaP, Phenols in smoke	100,000
GC/Ms	Nicotine, vOCs, Carbonyls, PAHs, flavouring compounds, aromatic amines in smoke	150,000
HPLC-MS/MS	TSNAs in tobacco/smoke	250,000
Atomic absorption spectroscopy	Metals in tobacco/smoke	50,000
Inductively coupled plasma–atomic emission spectroscopy	Metals in tobacco/smoke	70,000
Inductively coupled argon plasma–mass spectrometry	Metals in tobacco/smoke	200,000

Note: These costs are only approximate and may vary substantially depending on country-specific differences

Volatiles: benzene, 1,3-butadiene, acrylonitrile

Carbonyls: acrolein, formaldehyde, acetaldehyde, crotonaldehyde

Metals: arsenic, cadmium, chromium, lead, mercury, nickel, selenium

4.2 Accreditation

All analytical labs should be accredited by an international or national body. The standard for laboratories is ISO/IEC 17025. *(25)* This applies to all forms of testing laboratories: drug laboratories, environmental laboratories, tobacco product laboratories and others.

ISO 17025 addresses general lab competencies and management. It evaluates whether laboratories have the systems and protocols in place to document methods, staff qualifications and training, measurement verification and error minimization. It is broad enough to allow for laboratories that use standard methods, widely-accepted methods and laboratory-developed methods.

ISO 17025 does not and is not designed to evaluate whether methods used by a laboratory are accurate, reproducible and sensitive enough to make measurements fit for a particular application. For example, it is not intended to evaluate which analytical method is the most appropriate for a particular analysis. This is generally done through intra- and inter-laboratory verification. Thus accreditation is a necessary, but not sufficient factor in accessing laboratory competence.

Because of the general nature of the ISO standards, they are considered a minimum requirement for laboratories, but are not sufficient for demonstrating that a laboratory is able to provide accurate and reproducible analytical results. This is only proven through a complete quality assurance programme as described in Section 3.1.

4.3 Case study – Singapore

A good example of a laboratory that uses pre-existing government laboratory capabilities is the Cigarette Testing Laboratory (CTL) in Singapore. The CTL, together with the Pharmaceutical Laboratory and Cosmetics Laboratory, make up the Pharmaceutical Division at the Health Sciences Authority of Singapore. Established in the late 1980s, the CTL was tasked to test for tar and nicotine in mainstream cigarette smoke in support of tobacco regulatory compliance. It later expanded its scope to deal with toxicants beyond tar and nicotine by utilizing existing analytical facilities in the pharmaceutical and cosmetics laboratories. This approach allowed the laboratory to expand its capabilities at marginal additional cost.

Besides assisting capacity-building for other countries through training as part of the WHO TobLabNet, the laboratory also supports tobacco testing initiatives from countries requiring testing facilities to support their tobacco regulatory framework. These countries include: Fiji, Brunei, Tonga, the Solomon Islands and Samoa. This effort, which utilizes available testing laboratory facilities to build capacity and support tobacco regulatory compliance, provides a good model for other countries.

4.4 Step-by-step process

1. **Do your own internal research.**
    a. Identify a reasonable first set of analyses for your tobacco testing that would address your country's priorities.
    b. Identify the analytical requirements for the intended use of the data.

TOBACCO PRODUCT REGULATION
Building laboratory testing capacity

    c. Determine the estimated initial workload (number of samples over what period). This may be determined by the number of brands/subbrands being marketed and the frequency and schedule of testing.

    d. Identify the instrumentation (See Table 4) that is needed to carry out the analyses

2. **Discuss your approach with another country's regulatory agency that has experience with tobacco product testing.**
3. **Visit other government laboratories that are already doing consumer product testing.**
    a. Check the requirements listed above and rank available laboratories.
    b. Evaluate ability of laboratories to meet analytical accuracy, reproducibility, sensitivity and selectivity requirements to identify satisfactory laboratories.
    c. Use estimates of capacity requirements to identify acceptable laboratories.
4. **Negotiate with other government organizations, as appropriate, to obtain agreement to collaborate on testing.**
5. **Discuss particular needs and requirements for expected sample workload, turnaround time, reporting requirements, priority conflicts, etc. with the identified laboratory.**
6. **Finalize any required contractual agreement.**
7. **Communicate to the companies:**
    a. what needs to be tested and when?
    b. which laboratories are acceptable?

Chapter 5 Developing a Tobacco-exclusive Testing Laboratory

For the following discussion, an exclusive tobacco testing laboratory means a laboratory within a government system that does not share resources (equipment and personnel) with other programmes although it may be housed in the same physical facility. Developing an independent government tobacco testing laboratory can be a significant challenge unless starting with an existing laboratory capability, because the time and funds required can be considerable. It can also be challenging to maintain administrative support to see the project through to completion. Several countries have attempted to build independent laboratory facilities without expanding current capabilities, but to date these have been unsuccessful. Organizations that have been able to establish exclusive tobacco testing laboratories have typically built these capabilities on the foundations of another laboratory testing programme, to the point that they are self-sustaining and independent (see the example given in Section 5.4).

5.1 Requirements (infrastructure, laboratory equipment, staff, overall cost)

The facility footprint, equipment, and resources required depend on the expected scope of the testing programme. Making a clear thoughtful strategic determination of these requirements early in the process is a critical step and will greatly impact whether the entire programme is successful long-term. This cannot be overemphasized.

Organizations expecting to establish a laboratory with broad capabilities must anticipate the expected space requirement. Previously, TobReg gave recommendations for the facilities for a tobacco testing laboratory in 2004. *(1)* This document provides the following recommendations for a testing laboratory (See Table 5):

Table 5   Space requirements of a testing laboratory

Type of area/ accommodation	Minimum surface area (m²)	Expanded laboratory surface area[a] (m²)	Conditions
Preparation laboratory	20	60	Water and drainage required. Metals analysis will require a separate "clean room"
Smoking laboratory	20	60	Contains smoking machine(s). Air-conditioned and humidity-controlled (22 ± 2 °C and 60 ± 5%)
Instrument room	30	80	Air-conditioned; specialized instruments will require additional ventilation and other specific environmental controls
Offices	20	40	
Storage	15	25	
Common area	15	25	
Utility room	15	40	
Total	135	330	

[a] An expanded laboratory would include the necessary equipment for performing all recommended analyses of chemical constituents.

This space-requirement description is only an estimate based on what is typically expected for testing needs. A programme not intending to carry out as many analyses would need less space and a programme that intends

Chapter 5 Developing a Tobacco-exclusive Testing Laboratory

to perform more analyses would need more. It is highly recommended that, before programmes make final space decisions, they visit a currently operating tobacco testing laboratory to better understand the anticipated requirements.

The list provided in Table 4 identifies the basic equipment that may be needed for furnishing the analytical capabilities of the laboratory to carry out the analyses identified as a high priority. Additional equipment as described in section 4.1 may be needed for an expanded laboratory. In addition to the equipment listed in Table 4, standard laboratory equipment may be required.

A well-qualified and trained staff is necessary for successful tobacco product testing. For many of the analyses, specialized training will be necessary. As with floor space and equipment, the number of staff will depend on the expected number of analytical methods to be supported and the number of samples expected for analysis over the course of a year. In the same document cited above *(1)*, TobReg also provided an initial recommendation for staffing based on a typical laboratory performing testing on 150 brands/subbrands per year. TobReg recommended the following:

- one smoke laboratory manager;
- two-to-three smoke technicians, who should be familiar with the operation/maintenance of the smoking machine(s);
- two-to-three analytical chemists, who should have extensive knowledge of instrumentation; and,
- one quality control manager to supervise and control data and methods, and who should be well versed in statistics and data reporting.

The staffing should be increased proportionally if a higher sample load is expected. This is the minimum and does not include administrative staff or other non-technical staff who may be needed to provide support for laboratory operations.

5.2   Information technology (IT) systems

An effective and efficient IT system is critical for reducing errors and reporting results in a timely manner. The importance of IT systems is often overlooked by those unfamiliar with testing laboratory requirements. A significant part of background discussions with existing testing laboratories should include a discussion of the capabilities and requirements of current IT systems. An IT system for a laboratory should, at a minimum, include capabilities to:

- allow for logging in new samples
- enable scheduling of samples to be analysed based on changing priorities
- track samples through the analysis process
- allow for automatic data calculations where appropriate (most analytical equipment has internal systems that schedule, process, and assist analysts in analysing the raw data, but these systems must be compatible with the overall IT system)
- evaluate quality control results independent of the analyst
- reschedule analysis of samples that did not meet QC requirements
- report final results
- archive all data
- backup all data.

5.3 Data verification

Systems and processes must be put in place that allow for careful data verification before any data is reported. For experienced external laboratories or laboratories already in place for testing related to other regulatory programmes, systems should already be in place. For a newly established independent internal laboratory, systems will need to be developed.

5.3.1 Analysis of quality control materials

Quality control materials are samples introduced into every analytical run to ensure that systems are operating properly. There are well-established principles to assess quality control materials. *(26)* When results determined from the analyses of quality control materials deviate from a statistically acceptable range, results are rejected and investigations of the analytical systems are necessary.

5.3.2 Systematic checks of accuracy and reproducibility

Accuracy and reproducibility of analytical methods should be established before any method is used for analysis as described in Appendix 1. But changes in equipment, or other conditions, can cause these initial results to no longer be correct. Periodic checks of accuracy, by analysing known reference materials, and reproducibility, by performing duplicate analysis, should be performed to confirm that systems are continuing to operate within the original conditions and the initial measures are valid.

5.3.3 Long-term trend analysis

Long-term deviation of analytical results can be hard to identify because of the nature of these trends. Other quality assurance systems that are designed to monitor on a shorter time frame may not identify long-term deviation. Systems should be in place to ensure that long-term drift of data is identified and corrected.

5.4 Case study–CDC

In 1994, the Office on Smoking and Health (OSH) at US Centers for Disease Control and Prevention (CDC) approached staff of the laboratory of the National Center for Environmental Health (NCEH) to provide support in meeting certain regulatory requirements for review of information provided to OSH by the tobacco industry. Thus a clear mandate for testing was established. At that time, the laboratory already had a dedicated staff with 10 years of experience in using advanced analytical instrumentation applied to biomonitoring to evaluate exposure to users and non-users of tobacco products. Laboratory buildings were already in place with required environmental control and an uninterrupted power supply. In addition, service contracts for the equipment were in place with replacement parts on site to enable quick repairs. The laboratory already had an extensive quality control programme and statistical and IT support. Finally, there was a strong support structure in place that was not dependent on quick results but understood the need for a strategic approach.

Several staff from the NCEH laboratory visited the private commercial

Chapter 5 Developing a Tobacco-exclusive Testing Laboratory

tobacco analysis laboratory, Labstat Incorporated, in Kitchener, Canada. The staff graciously explained all of the requirements (environmental controls, equipment, staff, etc.) needed to outfit a successful tobacco testing laboratory. This allowed the staff of the NCEH laboratory to understand what other requirements were necessary to successfully develop the laboratory capabilities to test tobacco products. The laboratory purchased an environmental chamber, a smoking machine and particular tobacco product design testing equipment that were specific to tobacco analysis. This specialty equipment complemented the more general laboratory equipment needed to make analytical measurements. In a short time, the NCEH tobacco laboratory acquired equipment, laboratory space, and resources separate from the remainder of the NCEH laboratory so that equipment was no longer shared between programmes but was dedicated to tobacco product testing and research.

Several specific critical events, including the Philip Morris recall of 1995 *(27)*, occurred over the next few years that provided opportunities for the NCEH laboratory to demonstrate the value to the overall tobacco control programme at CDC. Since then, the laboratory has grown in size so that it has extensive capabilities and provides analytical support of the CDC mission and tobacco product research for the US FDA. The laboratory also serves as a training laboratory and works with FDA's own Southeast Regional Laboratory to develop and validate new methods for compliance testing.

The CDC tobacco laboratory was derived from capabilities that were already present. But the laboratory no longer shares equipment and personnel with other programmes. This process happened over a matter of years, allowing the laboratory to establish itself with only a minimal original investment. It also allowed the laboratory to grow as requirements and funding became available.

5.5 Step-by-step process

1. **Do your internal research.**
    a. Identify a reasonable first set of analyses for tobacco testing that would address your country's priorities.
    b. Identify the analytical requirements for the intended use of the data.
    c. Determine the estimated initial workload (number of samples over what period). This may be determined by the number of brands/subbrands being marketed and the frequency and schedule of testing.
    d. Identify the instrumentation (see Table 4) needed to carry out the analyses.
2. **Visit other laboratories that are already doing tobacco product testing to better understand the requirements of tobacco product testing.**
    a. Identify space, equipment and human resource needs.
    b. Identify specialty equipment and training that will be needed.
3. **Secure administrative assurance of long-term support and funding to create and sustain a laboratory capacity.**
4. **Work in close consultation with an established tobacco testing laboratory.**
    a. Develop laboratory facilities as needed.
    b. Hire experienced staff.
    c. Purchase analytical equipment
5. **Send staff for training at established laboratories.**
6. **Operationalize a limited number of laboratory methods above.**

Chapter 5 Developing a Tobacco-exclusive Testing Laboratory

7. Carry out intra-laboratory validation of methods by evaluating analytical accuracy, reproducibility, sensitivity and selectivity.
8. Participate in inter-laboratory validation exercises or exchange samples for analysis with experienced laboratories.
9. Expand capabilities by repeating 5–8 above.
10. **Communicate to the companies:**
    a. what needs to be tested and when?
    b. which laboratories are acceptable?

Chapter 6 Resources: WHO TobLabNet Membership (criteria, advantages, and procedures)

WHO TobLabNet laboratories can serve as a vital resource for any regulatory agency that is considering developing testing capabilities through any of the mechanisms described above. When considering how a laboratory can fit into a tobacco regulatory programme, visiting a WHO TobLabNet laboratory can be a valuable opportunity to see how other countries have approached this challenge.

WHO TobLabNet member laboratories can provide very important information about space requirements, environmental (electrical requirements, air conditioning, water quality, etc.) requirements, analytical instrumentation and staffing needs. By visiting a WHO TobLabNet member laboratory, government regulatory agents can see first-hand how laboratories are designed and discuss with WHO TobLabNet members their lessons learned and the steps that can lead to success.

Typically, the purchase of advanced analytical equipment includes some training by the manufacturer/supplier/vendor. This training usually consists of the basics of operation and maintenance of the instrument and using the software. But this training will not be specific enough to enable staff members to perform measurement of the design, contents and emissions of tobacco products. As time allows, WHO TobLabNet members can provide training for analysts in how to make tobacco product-specific measurements to complement any instrument manufacturer/ supplier/ vendor training.

Chapter 6 Resources: WHO TobLabNet Membership (criteria, advantages, and procedures)

To be most effective, training should take place after equipment has been purchased and installed and the analyst has had some hands-on operating experience. If an analyst is trained before equipment is installed, the training experience will be much less effective because they will not be able to ask more practical questions based on experience operating the instrument. Training may be done in the developing or newly established laboratory, or in a WHO TobLabNet laboratory.

The advantage of training in the developing/newly established laboratory is that whatever systems are set up will be present when the trainer leaves. But the analytical equipment must be installed and operational to make efficient use of the training time. Staff who take part in training should be the analysts who will actually be operating the equipment. Managers or officials who are not performing the day-to-day operation of analytical equipment are not appropriate for training. They will not be able to effectively communicate the lessons learned due to the technical nature of the information.

Requests for training can be made to staff of WHO TFI who can suggest laboratories that might be appropriate. WHO TFI is also developing a series of online training modules on the use of available WHO TobLabNet standard operating procedures to measure priority toxicants in cigarette tobacco filler and in mainstream cigarette smoke under ISO and intense smoking conditions. Further information will be available on the WHO TFI website as soon as the training platform is launched.

As an international laboratory network, WHO TobLabNet can serve as a valuable source of activities that help develop and demonstrate the abilities of member laboratories to make valid measurements. One example of this is the testing of a limited number of sample materials by two or more independent laboratories and/or two independent analytical methods. For example, if a laboratory is operationalizing a known method or developing a new method, carrying out measurements in more than one laboratory on shared samples can help demonstrate that accurate results are being reported.

In addition, inter-laboratory activities can be a valuable source of validation for participating laboratories. These include periodic round-robin exercises which are carried out either by direction of the WHO FCTC or developed as a work product by WHO TobLabNet. Previous round-robins have resulted in published documents which are available on the TFI website. *(28)*

WHO TobLabNet holds regular meetings of network members to encourage information exchange and planning for future joint projects. This can be valuable for new laboratories and can serve as a forum of addressing questions and comparing experience. The criteria for WHO TobLabNet laboratory membership are listed on the WHO TFI website *(28)* and are also listed below:

- the place the institution occupies in the country's health, scientific or educational structures;
- evidence of work in conjunction with the tobacco control community active within that country or geographic region;
- not be unduly influenced by relationships with organizations or entities with a significant financial stake in the outcome of the measurements;
- the quality of its scientific and technical leadership, and the number and qualifications of its staff;
- the institution's ability, capacity and readiness to contribute,

individually and within networks, to TobLabNet programme activities;
- experience with tobacco product testing or research or demonstrable intent to obtain capacity for tobacco product testing or research, e.g., commitment to train personnel and upgrade equipment;
- the institution's prospective stability in terms of personnel, activity and funding;
- the technical and geographical relevance of the institution and its activities to TobLabNet programme priorities;
- the working relationship which the institution has developed with other institutions in the country, as well as the at the inter-country, regional and global levels; and,
- the scientific and technical standing of the institution concerned at the national and international levels.

In addition, in order to prevent conflicts of interest, additional membership criteria *(29, 30)* are listed below:
- The laboratory should not be totally or partially owned by a tobacco company, however, laboratories that are owned or run by a national government that also owns or runs the national tobacco industry are allowed.
- Laboratories that receive funds from the tobacco industry in the form of feefor-service must demonstrate independence from the tobacco industry. For these organizations, a conflict of interest form is required.
- If a publicly-traded company, the tobacco industry should not have more than a 10% share of the total stocks.

- The laboratory should not have any member of the Board of Directors, or someone in a senior management position, who is employed by a tobacco company, which includes consultancy positions, among others. This also includes non-compensated consulting or advice to a tobacco company that may create a conflict by carrying the promise of future work.
- The laboratory may have tobacco companies as customers, but not its sole customers.

Summary

Tobacco product testing capability can be a valuable tool for countries trying to reduce the death and disease resulting from tobacco use by regulating the product. For the Parties to the WHO FCTC, the regulation of tobacco product contents and emissions (Article 9) and the regulation of tobacco product disclosures (Article 10) are among the key measures with the potential to contribute to reducing tobacco product demand. Although it is not an answer by itself, it can be used to inform and build on other regulatory activities, such as product review, product standards, packaging and labelling regulations, public education or as information to inform legislative decision-makers.

The approach to developing testing capabilities should be carefully considered strategically and based on clearly defined objectives. Those objectives will vary from country to country and can only be defined by considering individual national goals. This preliminary groundwork will pay off multiple times over by effectively using limited resources to achieve the maximum benefit.

While it may be possible to build a laboratory from the ground up, countries that have tried this strategy have not been successful. It is recommended that countries either contract with laboratories that are already testing tobacco products or build capabilities from existing laboratories with experience in testing other consumer products like pharmaceuticals, or environmental samples. This approach has been shown to be successful in several countries and provides the best

TOBACCO PRODUCT REGULATION
Building laboratory testing capacity

opportunity for accomplishing a country's objectives.

WHO TobLabNet was developed to support existing capabilities and to assist in developing new national tobacco product testing capabilities. There are numerous ways that WHO TobLabNet can help in laboratory development and countries interested in developing new laboratories should contact WHO TFI, who can put them in contact with appropriate WHO TobLabNet laboratories and coordinate activities to support the development of a testing programme. This contact should be made as early in the process of developing a laboratory as possible.

References

1. Guiding principles for the development of tobacco product research and testing capacity and proposed protocols for the initiation of tobacco product testing. Geneva, World Health Organization. 2004 (http://www.who.int/ tobacco/publications/prod_regulation/goa_principles/en/).
2. Brazil - Flavoured cigarettes banned. Geneva, WHO FCTC, March 2012 (http://www.who.int/fctc/implementation/ news/news_brazil/en/).
3. Canada (Legislative Summary of Bill C-32: An Act to amend the Tobacco Act), Publication Number LS-648E, 2009 (https://lop.parl.ca/About/Parliament/LegislativeSummaries/bills_ls.asp?ls=C32&Parl=40&Ses=2).
4. Canada (Order Amending the Schedule to the Tobacco Act (Menthol) P.C. 2017-256 March 24, 2017). (http://gazette. gc.ca/rp-pr/p2/2017/2017-04-05/html/sordors45-eng. php).
5. Regulation, the Tobacco Atlas. (http://www.tobaccoatlas.org/topic/regulations/, accessed 15 January 2018).
6. Brewer NT, Morgan JC, Baig SA, Mendel JR, Boynton MH, Pepper JK, et al. Public understanding of cigarette smoke constituents: three US surveys. Tob Control. 2016 Sep;26(5):592-599.
7. Morgan JC, Byron MJ, Baig SA, Stepanov I and Brewer NT. How people think about the chemicals in cigarette smoke: a systematic review. J Behav Med. 2017 Aug;40(4):553564).
8. Canada (Regulatory Impact Analysis Statement, Canada Gazette, Part I: Vol. 145 2011). http://www.gazette.gc.ca/ rp-pr/p1/2011/2011-02-19/html/reg3-eng.html
9. Kelley DE, Boynton MH, Noar SM, Morgan JC, Mendel JR, Ribisl KM et al. Effective Message Elements for Disclosures about Chemicals in Cigarette Smoke. Nicotine Tob Res. 2017 May 17. doi: 10.1093/ntr/ntx109. [Epub ahead of print].

10. Hecht SS. Approaches to cancer prevention based on an understanding of N-nitrosamine carcinogenesis. Proc Soc Exp Biol Med. 1997 Nov;216(2):181-91.
11. Hecht SS. Lung carcinogenesis by tobacco smoke. Int J Cancer. 2012 Dec 15;131(12):2724-32.
12. Ashley DL, O'Connor RJ, Bernert JT, Watson CH, Polzin GM, Jain RB et al. Effect of differing levels of tobacco-specific nitrosamines in cigarette smoke on the levels of biomarkers in smokers. Cancer Epidemiol Biomarkers Prev. 2010 Jun;19(6):1389-98.)
13. Stephens WE, Calder A, Newton J. Source and health implications of high toxic metal concentrations in illicit tobacco products. Environ Sci Technol. 2005; 39:479-88.
14. Tobacco, Euromonitor International (http://www.euromonitor.com/tobacco); International Tobacco Control Policy Evaluation Project (http://www.itcproject.org/resources/reports/2/); Latest, the Tobacco Atlas (http:// www.tobaccoatlas.org/); Tobacco Control Country profiles, WHO Tobacco Free Initiative. http://www.who.int/tobacco/surveillance/policy/country_profile/en/ All accessed 15 January 2018.
15. Canada. Tobacco Reporting Regulations. (https://www.canada.ca/en/health-canada/services/health-concerns/ tobacco/legislation/federal-regulations/tobacco-reporting-regulations.html, accessed 15 January 2018).
16. Brazil. Resolution – RDC No. 90, of December 27, 2007. (http://www.tobaccocontrollaws.org/files/live/Brazil/ Brazil%20-%20RDC%20No.%2090.pdf, accessed 15 January 2018).
17. Harmful and Potentially Harmful Constituents in Tobacco Products and Tobacco Smoke; Established List. A Notice by the Food and Drug Administration. 2012. (https://www. federalregister.gov/documents/2012/04/03/2012-7727/harmful-and-potentially-harmful-constituents-in-tobac-co-products-and-tobac-co-smoke-established-list).
18. Djordjevic MV, Fan J, Bush LP, Brunnemann KD, Hoffmann D. Effects of storage conditions on levels of TSNAs and N-nitrosamino acids in U.S. moist snuff. J. Ag-

ric. Food Chem. 1993; 41:1790-1794.
19. Andersen RA, Fleming PD, Hamilton-Kemp TR, Hildebrand DF. pH changes in smokeless tobacco undergoing nitrosation during prolonged storage: Effects of moisture, temperature, and duration. J. Agric. Food Chem. 1993; 41:968-972.
20. WHO Tobacco Laboratory Network (TobLabNet). WHO TFI, Geneva. http://www.who.int/tobacco/global_interaction/toblabnet/en.
21. WHO Tobacco Laboratory Network (TobLabNet) history. WHO TFI, Geneva. http://www.who.int/tobacco/global_interaction/toblabnet/history/en/index1.html
22. Best Practices in Tobacco Control - Regulation of Tobacco Products Canada Report, footnotes 3 and 4. WHO Study Group on Tobacco Product Regulation (TobReg). 2005. http://www.who.int/tobacco/global_interaction/tobreg/ Canada%20Best%20Practice%20Final_For%20Printing. pdf
23. Canada. Tobacco Reporting Regulations (SOR/2000-273), Justice Laws website, current to 27-12-11. (http://laws-lois.justice.gc.ca/eng/regulations/SOR-2000-273/page-2. html#h-2, accessed 15 January 2018).
24. The scientific basis of tobacco product regulation. Second report of a WHO study group. Geneva, 2008. (http://apps. who.int/iris/bitstream/10665/43997/1/TRS951_ eng. pdf?ua=1&ua=1).
25. ISO/IEC 17025:2005. General requirements for the competence of testing and calibration laboratories. International Organization for Standardization, Geneva. (https:// www.iso.org/standard/39883.html).
26. Westgard JO. A Total Quality-Control Plan with RightSized Statistical Quality-Control. Clin Lab Med. 2017 Mar;37(1):137-150.
27. Collins G. Tobacco giant recalls 8 billion faulty cigarettes. New York Times. 27 May 1995. http://www.nytimes.com/1995/05/27/us/tobacco-giant-recalls-8-billion-faulty-cigarettes.html.
28. Publications, WHO Tobacco Free Initiative website. http:// www.who.int/tobacco/publications/en/.
29. Designation, WHO Tobacco Free Initiative website. http:// www.who.int/tobacco/

global_interaction/toblabnet/designation/en/.
30. Participation and conflict of interest requirements, WHO Tobacco Free Initiative website. http://www.who.int/tobacco/global_interaction/toblabnet/conflict_of_interest/ en/.

Appendix 1. Intra- and Inter-laboratory Validation

A1.1 Intra-laboratory method validation

All laboratory methods must be carefully tested and a determination made that they meet requirements for how the data is to be used. Because different uses may have different requirements and therefore different equipment needs, the data requirements should be assessed before the laboratory decision process.

Accuracy is the nearness of a measurement of a quantity to the quantity's true value. Accuracy is primarily impacted by systematic error or bias. An analytical measurement may have an analytical bias and thus the result determined may be higher or lower than the actual true value. Accuracy is typically assessed by either testing the agreement with levels of materials with known values or testing the closeness among various testing regimens that are independent and should not have the same bias. In either of these cases, the closer the values that are determined are to the known level or consensus level, the more accurate the particular measurement under consideration. A lack of accuracy cannot be overcome by taking more measurements. In a common analogy, if a series of arrows were shot at a target, accuracy would be the closeness of the average of the various arrows to the centre of the target.

Precision is a determination of how close measurement results are to each other if a measurement is made repeatedly on the same sample, typically using the same method. Precision is primarily influenced by random error which causes the results to be inconsistent. Precision is typically assessed by making multiple measurements of the same material and then statistically determining the variability of the results compared to each other. Because precision is primarily determined by random error, the impact of a lack of precision on the accuracy of a measurement can be addressed, to some extent, by taking more measurements. The more measurements that are taken, the closer the average of these measurements will be to the value determined using the method. But calculated precision can be influenced by which steps in the method are included in the precision determination process. For example, results determined using replicate instrument analysis of the same sample after the sample preparation steps are completed will typically be more reproducible than data determined on samples that pass through both the sample preparation steps and the analytical measurement process; both parts of the analytical method can introduce random error. In the same analogy above, if a series of arrows are shot at a target, precision would be the closeness of the various arrows to each other even if the average taken together are not close to the centre of the target.

Sensitivity is the ability of a measurement to make accurate and precise determinations at low levels. Stated another way, it is the ability of an analytical system to detect an analyte if it is present. Sensitivity is impacted by the entire analytical process including analyte extraction, clean up, concentration, and analysis. Sensitivity can be represented by the limit of detection which is typically defined as three times the standard deviation of repeated measurement of a blank sample. Alternatively, it can be described by the limit of quantification which is defined at 10 times the standard deviation of repeated measurement of a blank sample. Sensitivity may be enhanced by improvements in any or all of the analytical steps; advances in instrumentation can provide substantial improvements.

A companion concept to sensitivity is selectivity or as it is also known, specificity. Selectivity is the ability to correctly identify that a substance is not present when it is indeed not present. Selectivity is primarily impacted by the presence of contaminants in a sample that have properties that are close enough to the analyte of interest to not be distinguished from the analyte of interest. Selectivity can be typically improved by better sample preparation methods and by more advanced instrumentation.

Ruggedness is the ability of an analytical system to withstand deviations from the defined analytical method. Deviations can include a wide range of phenomenon from errors in weighing materials to changes in instrument operation from one maintenance action to the next. For a proper assessment of ruggedness, the most likely deviations should be assessed in order to understand how these deviations will impact the final results. A proper ruggedness evaluation will identify those aspects that have the most impact on the measurement and should be most closely monitored.

A1.2   Inter-laboratory method validation

Inter-laboratory validation is important if data from one laboratory is to be compared with data from another. Also, if a laboratory wants to establish that their results agree with results that have been determined by others, an inter-laboratory validation is essential. There are several available programmes for assessment of inter-laboratory validation. A widely-accepted approach to assessing the data is found in ISO 5725-1 and ISO 5725-2. *(1, 2)* The inter-laboratory validation process consists of a single source providing equivalent samples to a series of laboratories. These samples are analysed using individual methods under the operating conditions in each laboratory and results reported back. All results are then evaluated to determine the repeatability and reproducibility of the results. By definition, the difference between two single results found for matched cigarette samples by the same operator using the same apparatus within the shortest feasible time will exceed the repeatability, r, on average not more than once in 20 cases in the normal, correct application of the method. Single results for matched cigarette samples reported by two laboratories will differ by more than the reproducibility, R, on average no more than once in 20 cases with normal, correct application of the method. The Cooperation Centre for Scientific Research Relative to Tobacco (CORESTA) has carried out several inter-laboratory validations to support efforts of the tobacco industry. *(3, 4)*. WHO TobLabNet has also carried out a series of inter-laboratory validation which are described in section 3.2.

References

1. ISO 5725-1:1994. Accuracy (trueness and precision) of measurement methods and results -- Part 1: General principles and definitions. International Organization for Standardization, Geneva. (https://www.iso.org/standard/11833.html).
2. ISO 5725-2:1994. Accuracy (trueness and precision) of measurement methods and results -- Part 2: Basic method for the determination of repeatability and reproducibility of a standard measurement method. International Organization for

Appendix 1. Intra- and Inter-laboratory Validation

Standardization, Geneva. (https://www.iso. org/standard/11834.html).
3. Intorp M, Purkis, S, Wagstaff, W, 2010a. Determination of aromatic amines in cigarette mainstream smoke: the CORESTA 2007 Joint Experiment; Beiträge Tabakforsch. Int. 24(2) (2010) 78-92.
4. Intorp M, Purkis, S, 2010b. Determination of selected volatiles in cigarette mainstream smoke. The CORESTA 2008 Joint Experiment; Beiträge Tabakforsch. Int. 24(4) (2014) 174-186.)